MARS

Readying rocket for liftoff from Mars, painted in 1956 by space artist Chesley Bonestell

Stunning vistas await the first humans on Mars.

MARS

OUR FUTURE ON THE RED PLANET

LEONARD DAVID

FOREWORD BY RON HOWARD

NATIONAL GEOGRAPHIC
WASHINGTON, D.C.

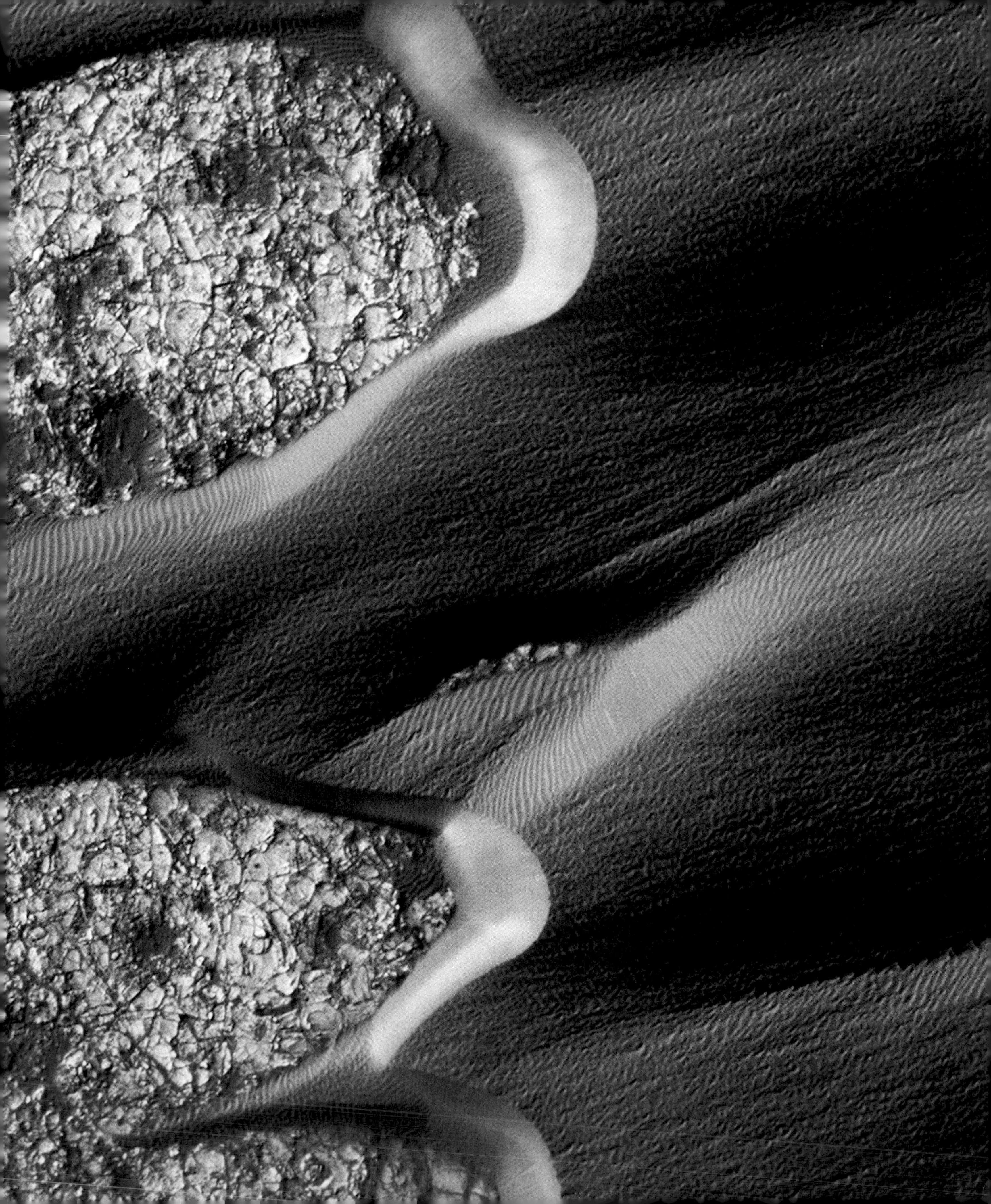

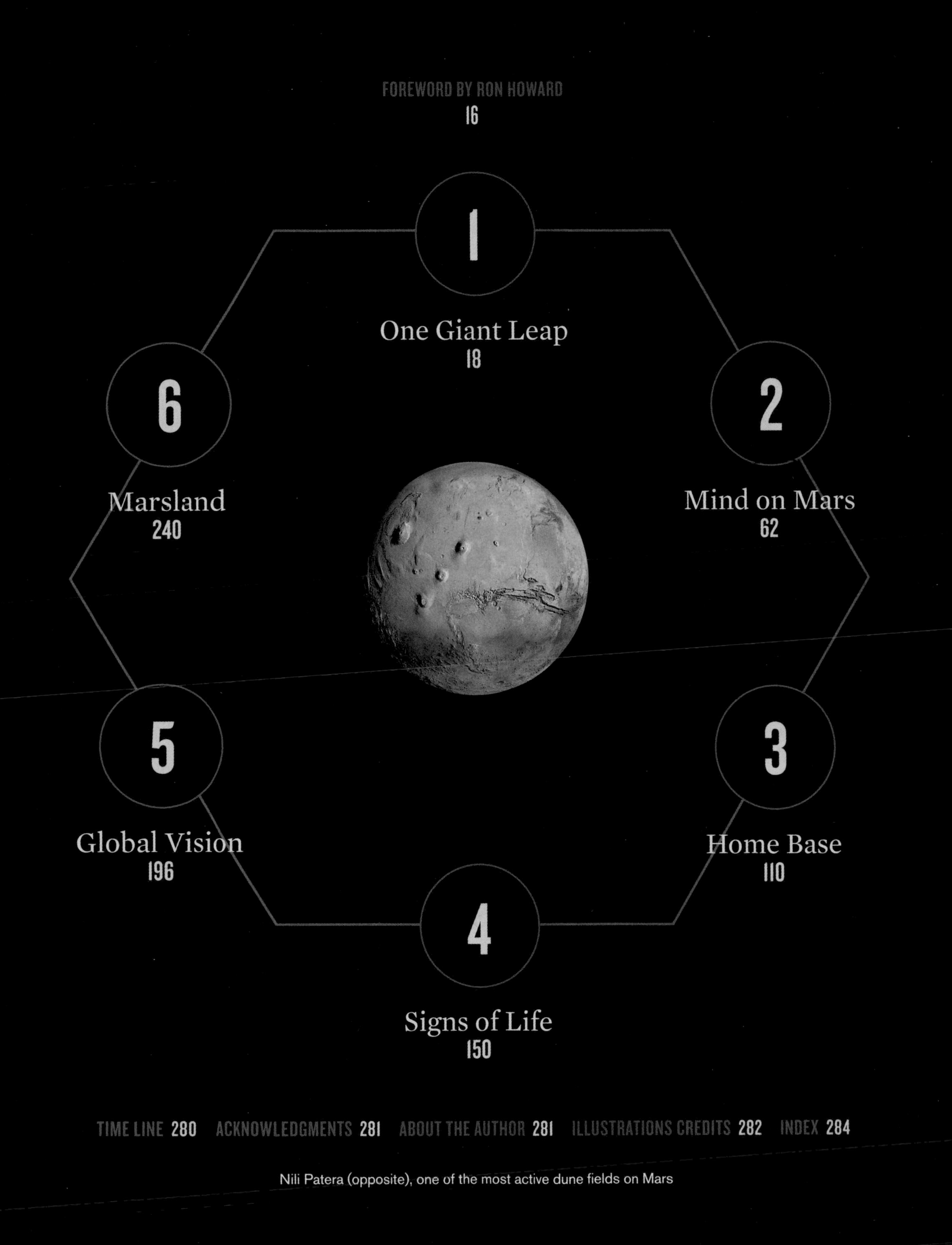

FOREWORD BY RON HOWARD 16

1 One Giant Leap 18

2 Mind on Mars 62

3 Home Base 110

4 Signs of Life 150

5 Global Vision 196

6 Marsland 240

TIME LINE 280 ACKNOWLEDGMENTS 281 ABOUT THE AUTHOR 281 ILLUSTRATIONS CREDITS 282 INDEX 284

Nili Patera (opposite), one of the most active dune fields on Mars

MARS EASTERN HEMISPHERE NORTH

This color mosaic is constructed from thousands of images returned from NASA's Mars Global Surveyor. This quadrant of Mars includes the site of the U.S. Viking 2 lander that touched down in September 1976.

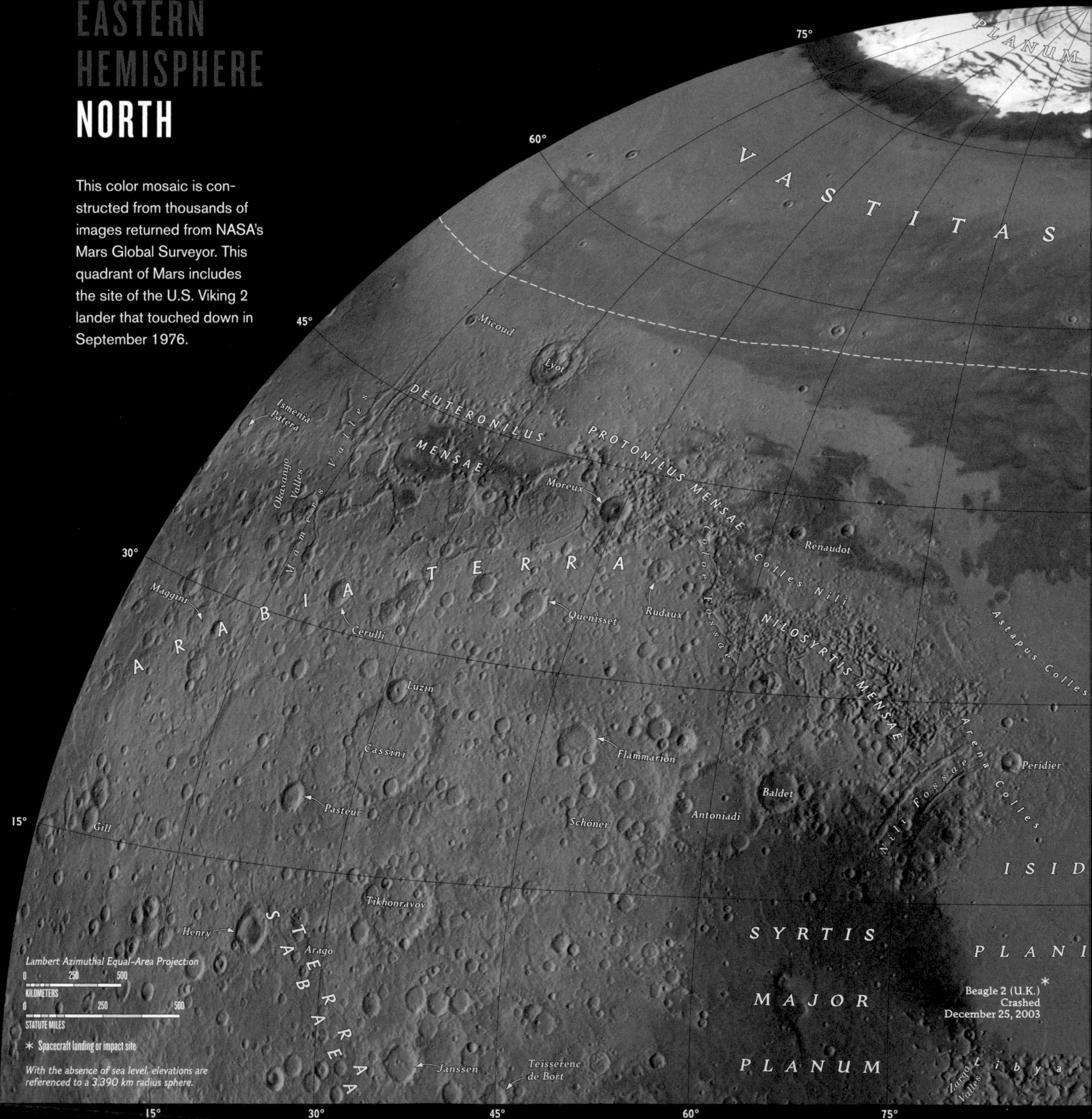

Lambert Azimuthal Equal-Area Projection

0 250 500 KILOMETERS

0 250 500 STATUTE MILES

* Spacecraft landing or impact site

With the absence of sea level, elevations are referenced to a 3,390 km radius sphere.

VASTITAS

PLANUM

DEUTERONILUS MENSAE

PROTONILUS MENSAE

NILOSYRTIS MENSAE

ARABIA TERRA

TERRA SABAEA

SYRTIS MAJOR PLANUM

ISID PLANI

Micoud

Lyot

Ismenia Patera

Okavango Valles

Mamers Valles

Moreux

Renaudot

Coloe Fossae

Colles Nili

Astapus Colles

Maggini

Cerulli

Quenisset

Rudaux

Luzin

Cassini

Flammarion

Arena Colles

Peridier

Nili Fossae

Baldet

Antoniadi

Schöner

Pasteur

Gill

Tikhonravov

Henry

Arago

Janssen

Teisserenc de Bort

Beagle 2 (U.K.) Crashed December 25, 2003

Zarqa Valles

Libya

15° 30° 45° 60° 75°

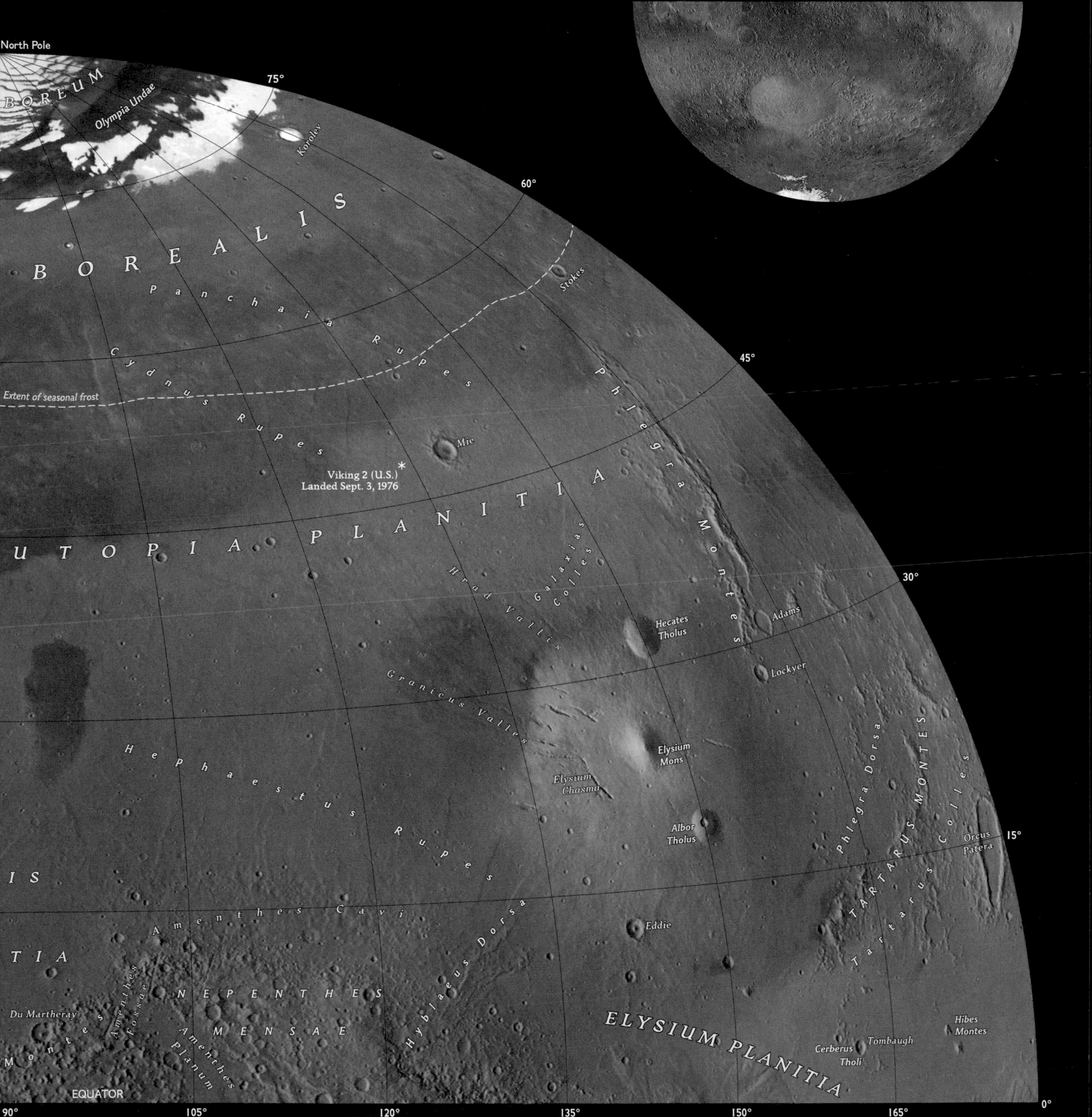
North Pole
BOREUM
Olympia Undae
75°
Korolev
60°
BOREALIS
Panchaia Rupes
Stokes
Cydnus Rupes
45°
Extent of seasonal frost
Phlegra Montes
Mie
Viking 2 (U.S.)
Landed Sept. 3, 1976
UTOPIA PLANITIA
Hrad Vallis
Galaxias Colles
30°
Hecates Tholus
Adams
Lockyer
Granicus Valles
Elysium Mons
Hephaestus Rupes
Elysium Chasma
Phlegra Dorsa
TARTARUS MONTES
Tartarus Colles
Albor Tholus
Orcus Patera
15°
IS
Amenthes Cavi
Eddie
Hyblaeus Dorsa
TIA
NEPENTHES
MENSAE
Amenthes Fossae
Du Martheray
Montes
Amenthes Planum
ELYSIUM PLANITIA
Hibes Montes
Tombaugh
Cerberus Tholi
EQUATOR
90°
105°
120°
135°
150°
165°
0°

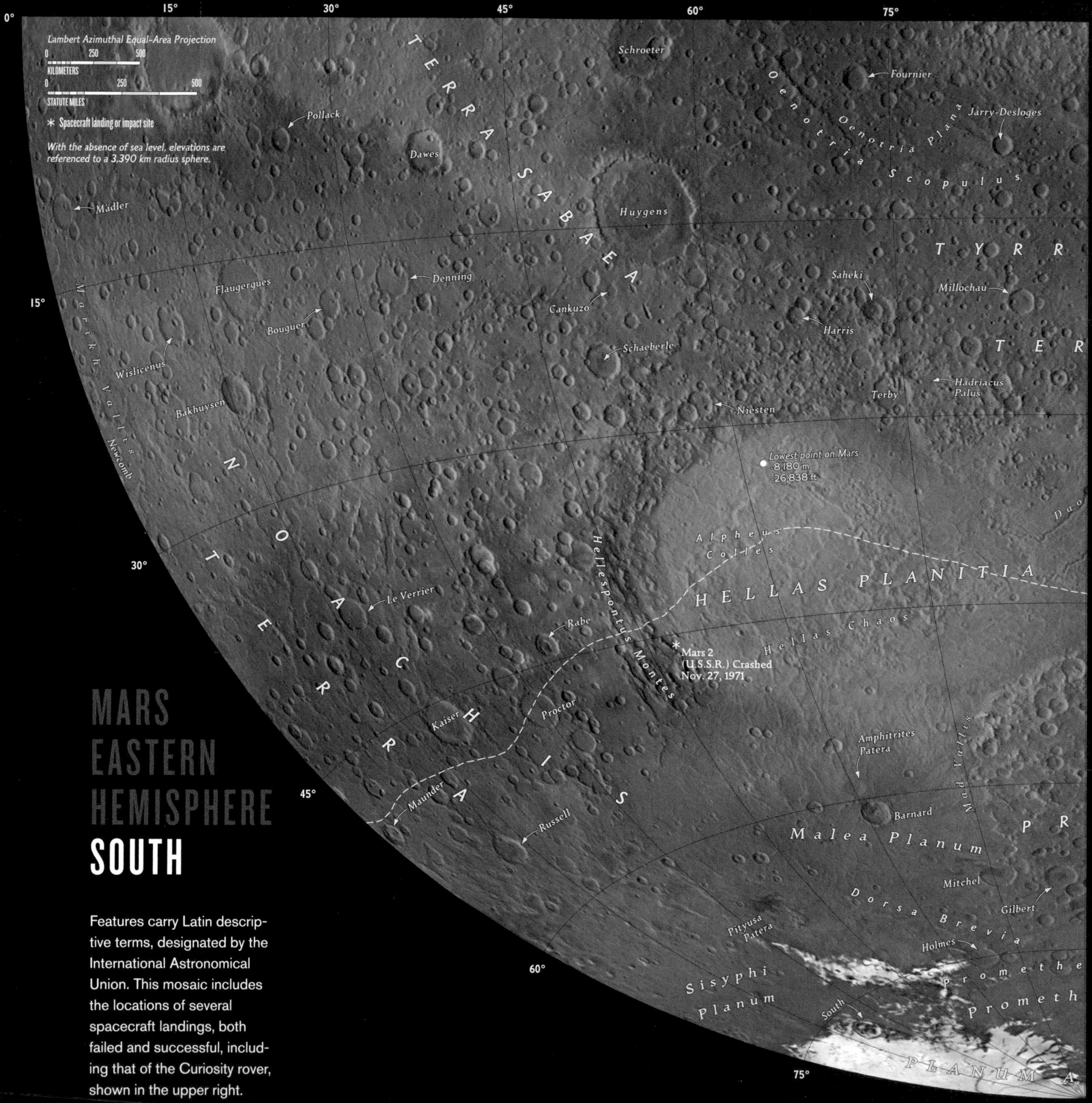

MARS EASTERN HEMISPHERE SOUTH

Features carry Latin descriptive terms, designated by the International Astronomical Union. This mosaic includes the locations of several spacecraft landings, both failed and successful, including that of the Curiosity rover, shown in the upper right.

Lambert Azimuthal Equal-Area Projection

0 250 500 KILOMETERS

0 250 500 STATUTE MILES

* Spacecraft landing or impact site

With the absence of sea level, elevations are referenced to a 3,390 km radius sphere.

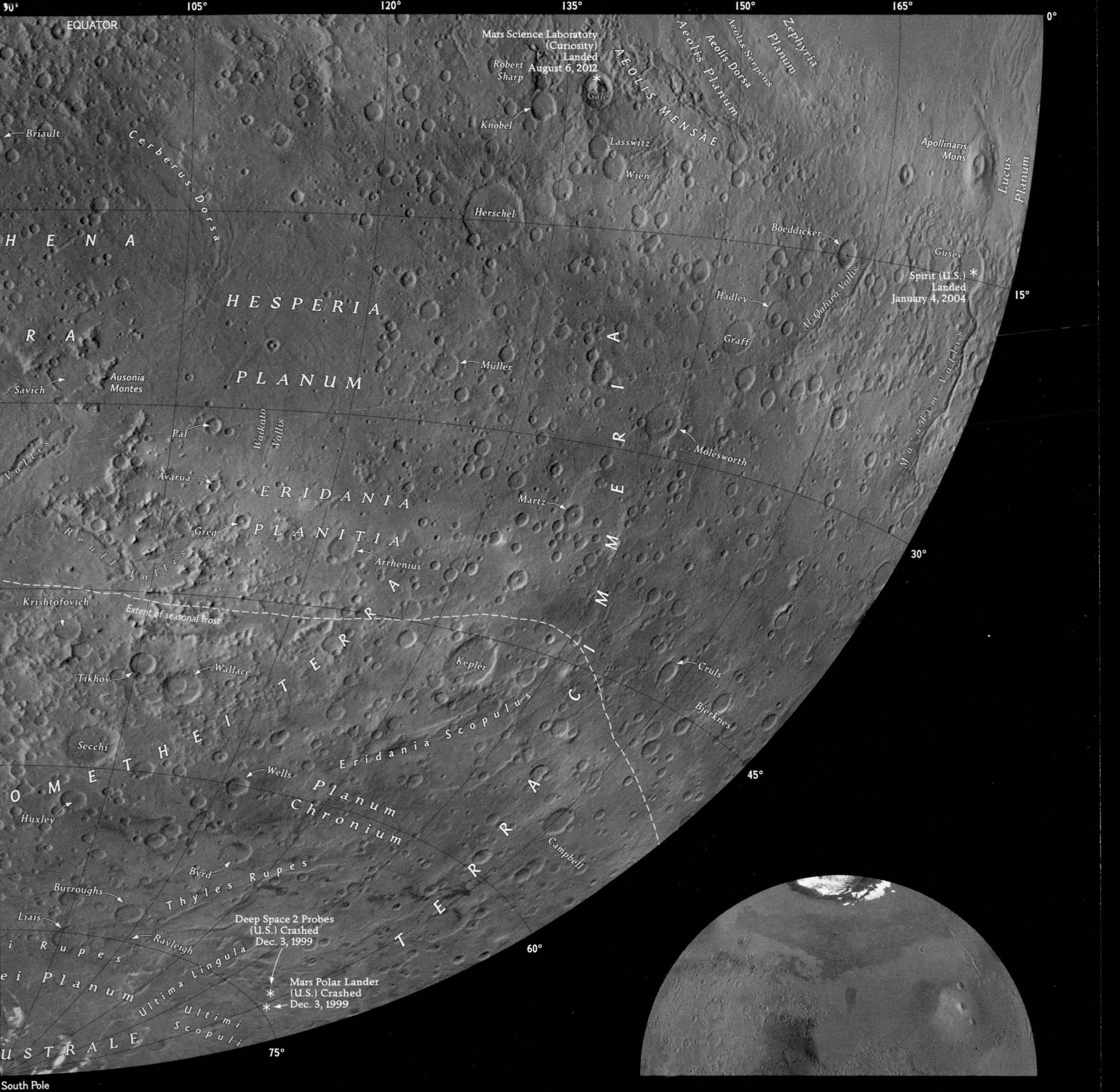
EQUATOR
105°
120°
135°
150°
165°
0°
15°
30°
45°
60°
75°
Mars Science Laboratory (Curiosity) Landed August 6, 2012
Robert Sharp
Gale
Knobel
Lasswitz
Wien
AEOLIS MENSAE
Aeolis Planum
Aeolis Dorsa
Aeolis Serpens
Zephyria Planum
Apollinaris Mons
Lucus Planum
Briault
Cerberus Dorsa
Herschel
Boeddicker
Gusev
Spirit (U.S.) Landed January 4, 2004
HESPERIA
PLANUM
Hadley
Al-Qahira Vallis
Graff
Müller
Ma'adim Vallis
Savich
Ausonia Montes
Pál
Waikato Vallis
Molesworth
Avarua
ERIDANIA
PLANITIA
Martz
Greg
Arrhenius
Reull Vallis
TERRA CIMMERIA
Krishtofovich
Extent of seasonal frost
Kepler
Cruls
Tikhov
Wallace
Bjerknes
Secchi
Eridania Scopulus
Wells
Planum Chronium
Huxley
Campbell
Byrd
Thyles Rupes
Burroughs
Liais
Rayleigh
Deep Space 2 Probes (U.S.) Crashed Dec. 3, 1999
Mars Polar Lander (U.S.) Crashed Dec. 3, 1999
Ultima Lingula
Ultimi Scopuli
South Pole

Windswept dust curves around a hill in Mars's Ganges Chasma.

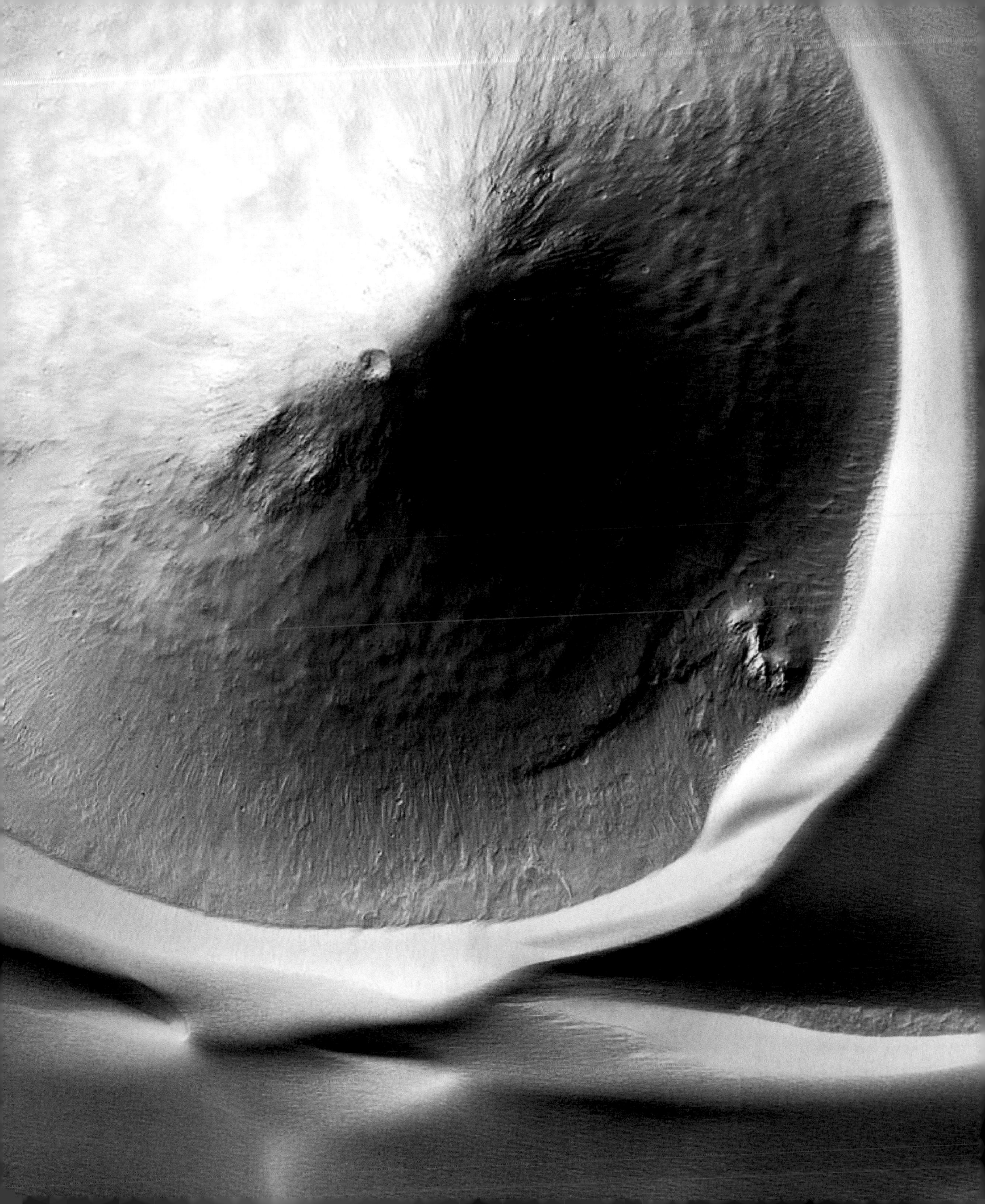

MARS WESTERN HEMISPHERE NORTH

The magnificent Olympus Mons, the largest of the volcanoes in the Tharsis Montes region, is found in this quadrant of the planet. Volcanoes in this area are ten to 100 times larger than those on Earth.

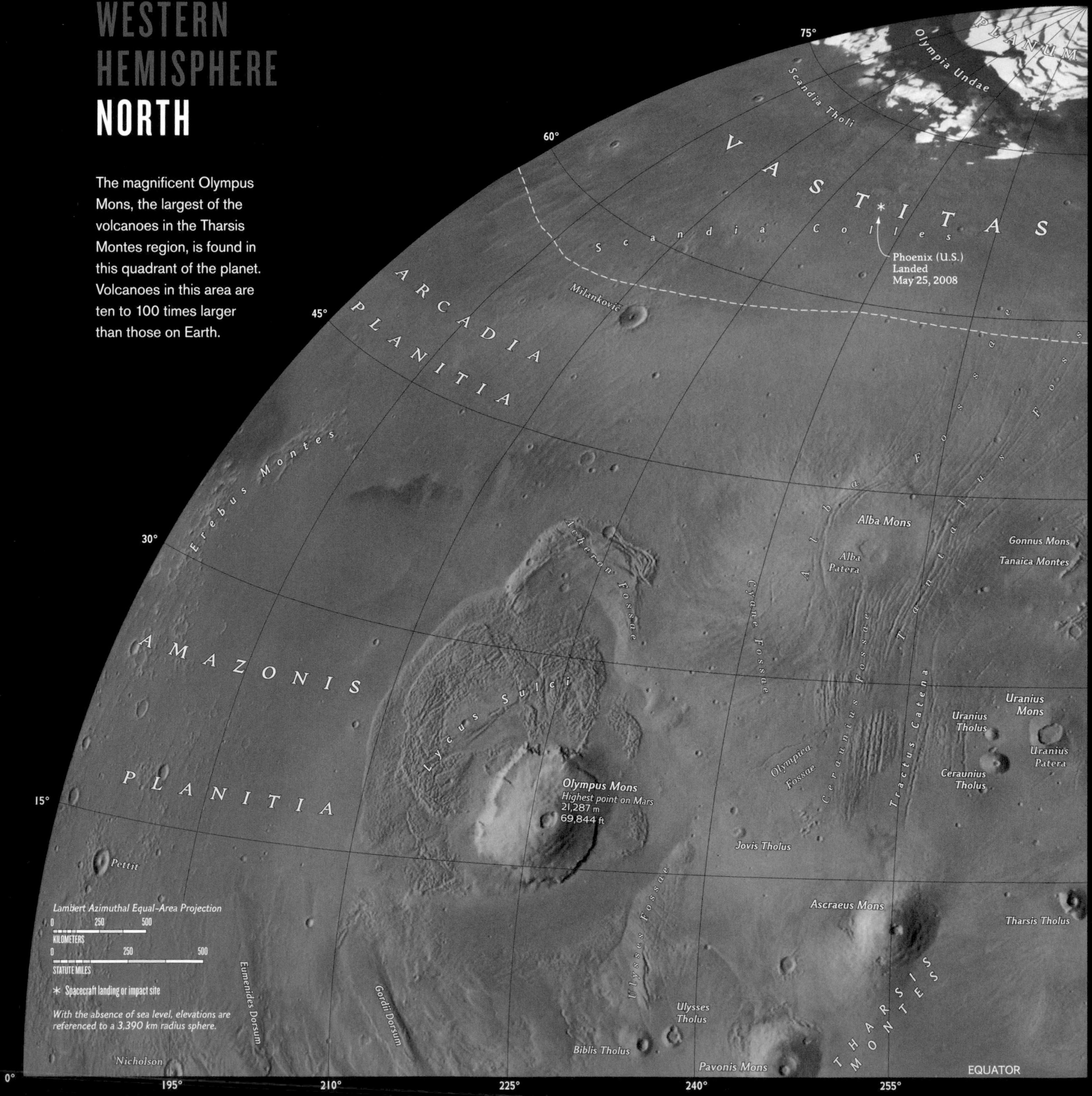

PLANUM
Olympia Undae
Scandia Tholi
VASTITAS
Scandia Colles
Phoenix (U.S.) Landed May 25, 2008
Milankovič
ARCADIA PLANITIA
Erebus Montes
Alba Fossae
Tantalus Fossae
Alba Mons
Alba Patera
Gonnus Mons
Tanaica Montes
Acheron Fossae
Cyane Fossae
AMAZONIS PLANITIA
Lycus Sulci
Ceraunius Fossae
Tractus Catena
Uranius Tholus
Uranius Mons
Uranius Patera
Ceraunius Tholus
Olympica Fossae
Olympus Mons
Highest point on Mars
21,287 m
69,844 ft
Jovis Tholus
Pettit
Ascraeus Mons
Tharsis Tholus
Ulysses Fossae
Ulysses Tholus
Biblis Tholus
Eumenides Dorsum
Gordii Dorsum
Nicholson
Pavonis Mons
THARSIS MONTES
EQUATOR

75° 60° 45° 30° 15° 0°
195° 210° 225° 240° 255°

Lambert Azimuthal Equal-Area Projection
0 250 500 KILOMETERS
0 250 500 STATUTE MILES
* Spacecraft landing or impact site

With the absence of sea level, elevations are referenced to a 3,390 km radius sphere.

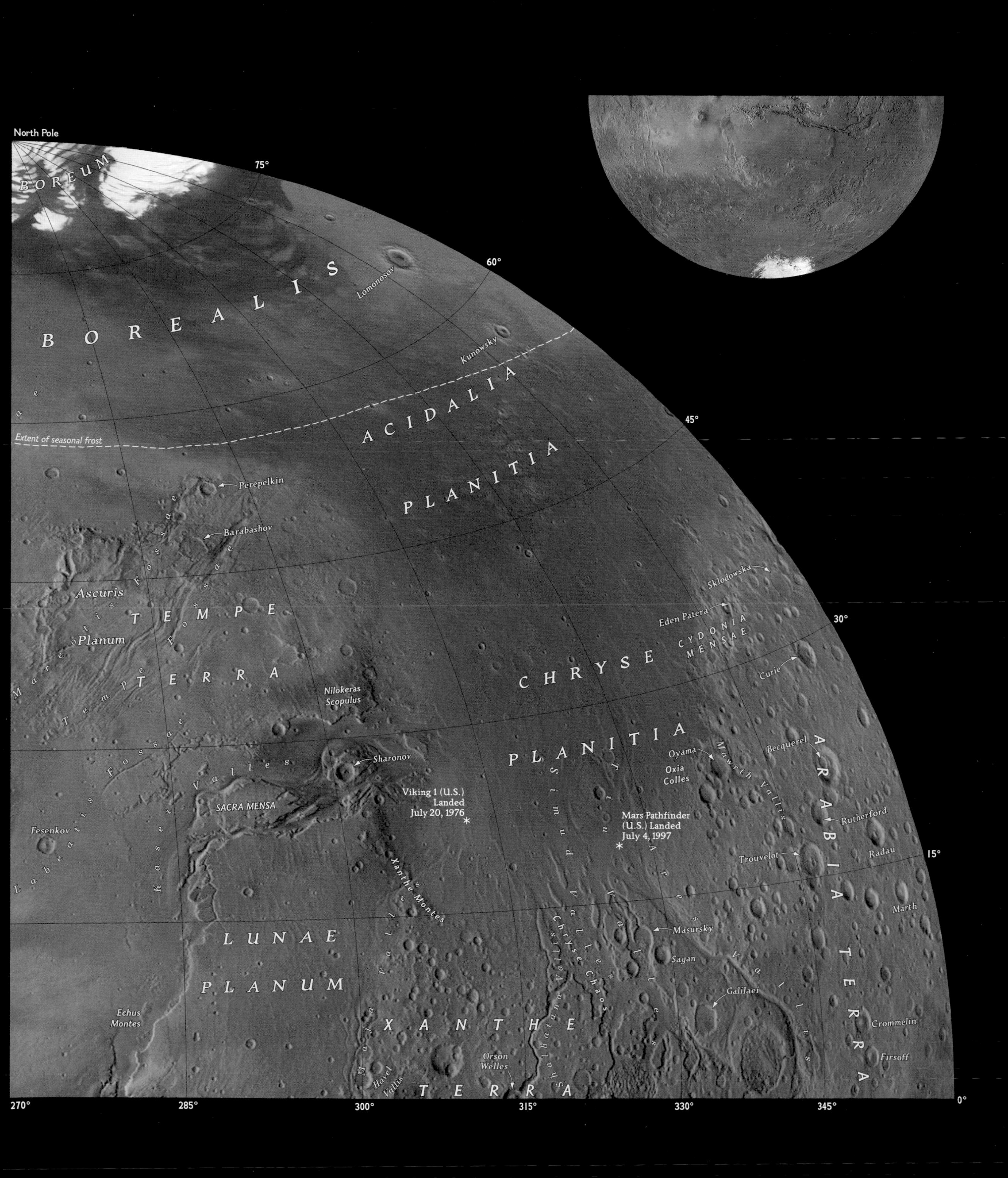
North Pole
75°
60°
45°
30°
15°
0°
270°
285°
300°
315°
330°
345°
BOREUM
BOREALIS
Lomonosov
Kunowsky
ACIDALIA PLANITIA
Extent of seasonal frost
Perepelkin
Barabashov
Tempe Fossae
Ascuris Planum
TEMPE TERRA
Mareotis Fossae
Tempe Fossae
Eden Patera
Sklodowska
CYDONIA MENSAE
CHRYSE PLANITIA
Curie
Nilokeras Scopulus
Sharonov
Kasei Valles
SACRA MENSA
Labeatis Fossae
Fesenkov
Viking 1 (U.S.) Landed July 20, 1976
Mars Pathfinder (U.S.) Landed July 4, 1997
Oyama
Oxia Colles
Mawrth Vallis
Becquerel
ARABIA TERRA
Rutherford
Trouvelot
Radau
Marth
Xanthe Montes
Simud Valles
Tiu Valles
Ares Vallis
Masursky
Sagan
Chryse Chaos
Galilaei
LUNAE PLANUM
Echus Montes
XANTHE TERRA
Maja Valles
Havel Vallis
Orson Welles
Shalbatana Vallis
Crommelin
Firsoff

MARS WESTERN HEMISPHERE SOUTH

This quadrant of Mars includes Valles Marineris, a grand system of canyons that stretches for nearly a quarter of the planet's circumference. The entire system extends over 2,490 miles.

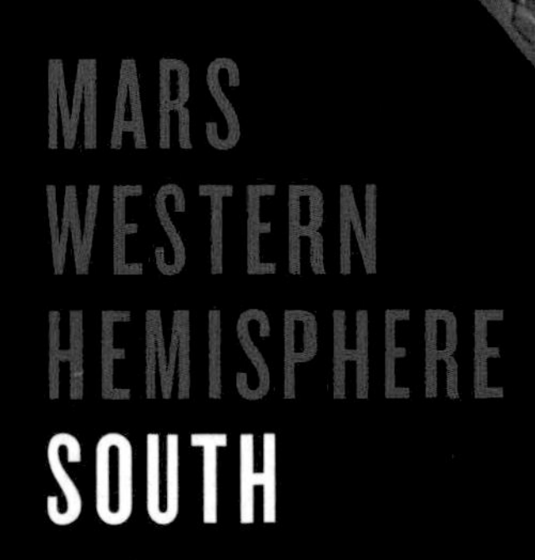

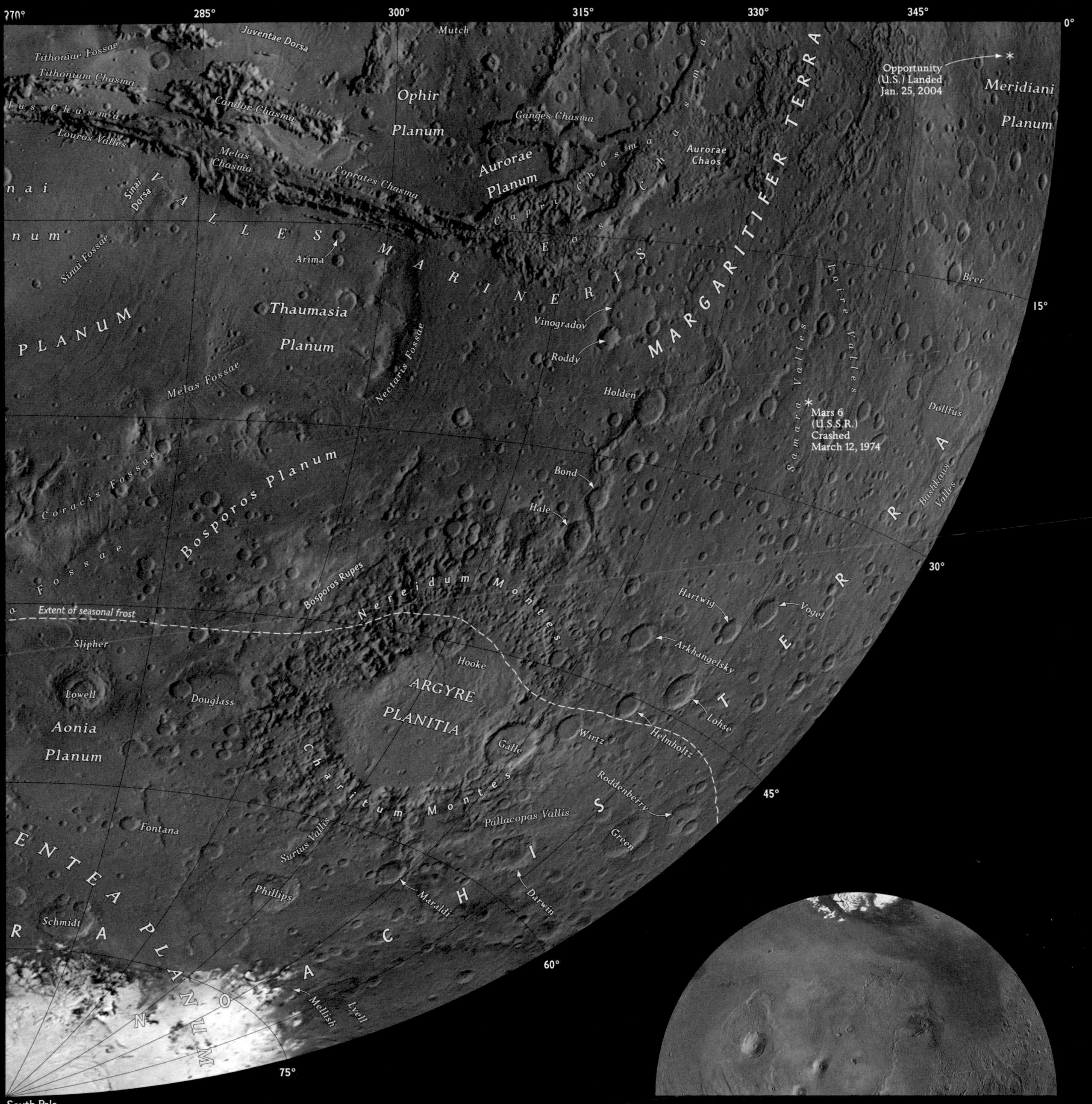

270°
285°
300°
315°
330°
345°
0°
15°
30°
45°
60°
75°
South Pole
Juventae Dorsa
Mutch
Tithoniae Fossae
Tithonium Chasma
Candor Chasma
Louros Valles
Melas Chasma
Coprates Chasma
Ophir Planum
Ganges Chasma
Aurorae Planum
Aurorae Chaos
Capri Chasma
Eos Chasma
Sinai Dorsa
VALLES MARINERIS
MARGARITIFER TERRA
Opportunity (U.S.) Landed Jan. 25, 2004
Meridiani Planum
Beer
Arima
Sinai Fossae
PLANUM
Thaumasia Planum
Vinogradov
Roddy
Nectaris Fossae
Melas Fossae
Holden
Samara Valles
Loire Valles
Mars 6 (U.S.S.R.) Crashed March 12, 1974
Dollfus
Bond
Hale
Bashkaus Valles
Bosporos Planum
Coracis Fossae
Bosporos Rupes
Nereidum Montes
Extent of seasonal frost
Hartwig
Vogel
Arkhangelsky
Slipher
Hooke
ARGYRE PLANITIA
Lowell
Douglass
Aonia Planum
Lohse
Helmholtz
Wirtz
Galle
Charitum Montes
Roddenberry
Pallacopas Vallis
Fontana
Surius Vallis
Green
Maraldi
Darwin
Phillips
Schmidt
Mellish
Lyell

Foreword

AS A BOY, I WAS EXCITED by the pioneer experience. Old movies like *How the West Was Won* sparked my imagination, as did stories of the European explorers who embarked on long and perilous ocean journeys in search of undiscovered lands.

Then, when I was 15 years old, we landed Apollo 11 on the moon. Along with hundreds of millions of other viewers, I watched with held breath as Neil and Buzz became the first humans to set foot on the lunar surface. I was deeply moved by the seeming impossibility of what we had just accomplished. During that broadcast, Mission Control relayed a phone call from President Nixon to the astronauts. He said to them, "Because of what you have done, the Heavens have become a part of man's world."

A new frontier had opened.

Just 26 years after that, I was making the film *Apollo 13*. I was thrilled at the chance to tell the personal stories of these brave explorers of the modern era who had risked their lives in the name of human potential. I had the remarkable experience of interviewing most of those early astronauts and many others who were involved in the space program. And every conversation pointed to one conclusion: We've come this far; we have to keep going. Buzz Aldrin was adamant when he told me that the next dream to work toward was putting humans on Mars.

In every generation, there are new frontiers—new places and ideas that we must discover, explore, and understand. That's curiosity—the force that drives us all and makes us human. We wonder. We ask questions. And in answering them, we learn and evolve.

We've been asking questions about sending humans to Mars for a long time. Science-fiction writers have been writing about it for over a century. It was just a matter of waiting for the moment when the technology could keep pace with our imaginations.

Visionaries like Elon Musk are saying that moment is now, which is why now is the perfect time to tell this story. When the *Mars* series project came to Brian Grazer and me, we were immediately excited by the vitality of the subject matter and the creative possibilities that the series offered. This is exactly the kind of storytelling challenge we've faced before and are always looking for—stories about the grandeur of the human spirit, told through the lens of the human hearts and minds that make it happen.

As the idea for the series was being developed, I began to think about Mars in ways that I never had before. The series would not just be about a mission to Mars—which had been addressed in movies and documentaries to some extent; it would take an even larger, longer, and more epic view into the actual colonization. And as I began to learn the staggering amount of research that had already been done into the viability of humans living on Mars, I was hooked. I had a new pioneer story to tell.

As the creative process and the team started coming together, we settled on a unique and original storytelling perspective: We would tell the story of colonizing Mars looking back from the future, with the premise that we've already gone to Mars, *and this is what it took to get us there.* To accomplish this, we created the series as a hybrid of documentary and scripted narrative, with our documentary footage serving as the past tense of the fictional future. It was an exciting challenge to combine these two genres in a new way that would transport audiences and give them a visceral—and realistic—sense of what it might be like to travel to and colonize Mars.

Every great mission requires a great team, and this project is no exception. I am grateful for all the partnerships that helped bring this to life: Brian Grazer and everyone at Imagine Entertainment; RadicalMedia, which is always masterful in executing its productions; and National Geographic, which has always been so dedicated to helping us all better understand the world we live in and worlds beyond. I am also very appreciative to National Geographic for being incredibly rigorous in its demand for authenticity and scientific accuracy. The level of vetting around the engineering and hard science throughout the scriptwriting process is something that we are proud of—and it's part of what makes this series unique.

This isn't science fiction at all. It's real science. And in the documentary aspect of the show, we have many—if not all—of the greatest minds in the world currently focused on this subject. They have acted as our trustworthy guides. It is my hope that this series and this book will serve as a unique historical record—something that people can look back on in decades to come and say, "They didn't get it all right, but look how much they already knew." That's my goal: to surprise people 50 years from now with the clear sense we had of what it would take—on the human level *and* the scientific level—to go to Mars and create a new civilization.

I hope this book fires people's imaginations, sheds light on the power and possibility of this unique historical moment, and inspires the next generation of pioneers.

I am honored to have played a part in bringing this vision into the world.

—Ron Howard
Director and producer, Imagine Entertainment

Editor's Note: This book is created in tandem with National Geographic's special television series, Mars, *produced by Imagine Entertainment and RadicalMedia. Each chapter harmonizes with an episode of the series, offering a brief plot summary and a fascinating in-depth look at the science, engineering, and ethical challenges facing us as we reach, explore, and inhabit Mars.*

ONE GIANT LEAP

Getting to Mars is the first challenge. After safe entry, descent, and landing on the planet's surface, humans arrive at a place no one has ever before called home.

Taking the plunge: An artist's concept of NASA's Mars Science Laboratory spacecraft approaching Mars, the Curiosity rover tucked inside. *Preceding pages (background):* Mars is a halo of scattered light, viewed by the European Space Agency's Rosetta spacecraft in 2006.

1

One Giant Leap

N THE 2030S, a peculiar shadow slips across the reddish vista that is Mars.

The historic arrival of the first expeditionary crew from Earth to the Red Planet balances on an impulsive mix of rocket propulsion, determined will, and hard-earned luck. Outstretched landing legs of the craft near the planet. A powered and safe descent by humans onto Mars is a literal trial by fire.

Robotic spacecraft have been there before. Over the previous decades, Mars has been flown by, circled, crashed into, pinged by radar, camera profiled, listened to, parachuted onto, as well as bounced and rolled over, shoveled and drilled into, smelled, baked, tasted, and laser zapped.

But to date there remained one missing element in the exploration of Mars: actually stepping on the planet. In the 21st century, the sandy face of that faraway world is to be dotted by the first footprints of humans.

The roaring engines of the lander are reduced in power, and the vehicle comes to a full stop, ending a voyage from Earth of millions of miles and several hundred days. The crew has endured the physical, psychological, and social stresses of long-duration space travel. But as the first crew members prepare to set boot on Mars, the human journey ahead, while momentous, is treacherous. As sojourners from Earth representing multiple nations take those early first steps, the third and fourth planets of the solar system may well be everlastingly linked. *Setting foot* on Mars is one thing, however. *Staying put* on Mars is far more daunting.

Getting humans off planet Earth and transiting to planet Mars means tossing into space lots of throw weight made feasible by heavy-lifting rockets. There's a lot of work involved in flinging humans and their accoutrements outward for the months-long journey: food, water, and exercise gear, not to mention radiation shielding and the essential supplies to haul to Mars.

Ready for touchdown! Artist's concept depicts NASA's Phoenix Mars lander using its pulsed rocket engines to make a 2008 touchdown on the arctic plains of Mars.

EPISODE I

NOVO MUNDO

Spacecraft *Daedalus,* carrying the first human mission to Mars, finally reaches its destination after a long journey. The world watches and waits to see if the ship will land safely, but as it enters the Martian atmosphere a problem with the thrusters is detected. Disaster is averted by the timely actions of the flight commander, but the rough descent leaves him injured. To further complicate matters, *Daedalus* has landed significantly off course from its intended target.

Scientists and engineers have begun to chart the ideal location for establishing the first human outpost on the Red Planet. That "best" site would not be selected solely for safety's sake. The place of choice must be of high scientific merit too, arguably, supplying the best conditions for finding the answer to whether Mars has spawned a "second Genesis" of life.

Mars is also being eyed for its resources to help prolong the stay time of expeditionary crews on the planet. A new lineup of spacecraft is now in blueprint stage to scour Mars for underground pockets of ice. Powerful communication orbiters are also a must for messaging and video links between Earth and Mars. The distance between the two worlds leads to delayed dialogue, even at the speed of light.

Down and dirty

THE FIRST CREW ON MARS will benefit from all manner of robotic craft that preceded them into the orbit and onto the surface of the Red Planet. Now circling Mars, for instance, NASA's Mars Odyssey, the Mars Reconnaissance Orbiter, and the Mars Atmosphere and Volatile Evolution Mission (MAVEN) are part of an international armada of spacecraft probing the Red Planet, a flotilla that includes Europe's Mars Express and India's Mars Orbiter Mission, or MOM.

Mechanical predecessors await on the planet's surface as well. Getting automated machinery "down and dirty" on Mars has been a hit-and-miss tale of retro-rockets, parachutes, airbags, and a complicated contraption called a sky crane. A progression of triumphant U.S. robotic landers involves two Viking landers in 1976; the Pathfinder/Sojourner rover in 1997; rover twins—Spirit and Opportunity—in 2004; and the Phoenix lander in 2008. And in August 2012, the largest payload ever to reach Martian real estate was the NASA Mars Science Laboratory mission, successfully depositing the car-size Curiosity rover. In total, that Mars machinery had a mass of roughly 1,984 pounds.

Landing far larger payloads—crew-carrying spacecraft and life-sustaining habitats—safely on the planet demands use of still untried landing technologies. Here's the catch: Placing equipment and crew on Mars is more difficult than rocketing onto Earth's moon, Apollo style, due to the higher gravity of Mars and the presence of an atmosphere. Yes, the Mars atmosphere is very thin, but it is still a force to reckon with, causing major heating on an entering spacecraft. It's significant enough to threaten overheating but not thick enough to enable landing with only aerodynamic decelerators, especially for the larger masses needed to support human missions.

Studies indicate that the best tactic is to use aerodynamic decelerators until a Mars craft is heading for the planet at supersonic speeds. Then the aerodynamic system separates, and descent module rocket engines ignite so that the vehicle makes a powered descent above Mars, a final approach, and a relatively controlled landing.

Add in the human factor, and the challenge is even greater. For example, to parachute the first crew onto Mars—say, tucked inside a 40-ton lander—would require a parachute the size of the Rose Bowl. Huge inflatable aerodynamic decelerators and supersonic retropropulsion are now considered the technologies of choice to get people and paraphernalia on Mars.

In the zone

WHERE ON THE PLANET is the sweet spot for the first crew visiting Mars? Key criteria have been defined, and NASA has already started to pinpoint locations. In October 2015, the First Landing Site/Exploration Zone Workshop for Human Missions to the Surface of Mars was held at the Lunar and Planetary Institute in Houston, Texas.

Nearly 50 locations on Mars were proposed by researchers at the gathering, an event termed "historic" and a "turning point," by James Green, director of planetary science at NASA Headquarters in Washington, D.C. "It's really the start of making Mars real . . . identifying the real locations for us to be able to land, work and do our science," he told attendees.

The prospective landing sites must meet several preordained NASA guidelines. First, each outpost should be encircled by a more than 60-mile-wide exploration

"A new consensus is emerging in the scientific and policy communities around our vision, timetable, and plan for sending American astronauts to Mars in the 2030s. More and more of our neighbors are finding new meaning in the words 'Mars matters.' "

—Charles Bolden, Administrator of NASA

zone. Touchdown places ought to be within 50 degrees of latitude, north or south, of the Martian equator. They should be able to support a set of three to five landings, allowing four- to six-person crews to carry out duties through an expedition lasting about 500 sols, or Martian days.

Obviously a key essential for picking any place on Mars is safety in landing. An incoming crew must stay away from boulder fields, steep slopes, sand dunes, craters, and strong winds. Toss in the prevalence of Martian dust that could gum up space suits, equipment, and habitat air locks. Toppling over in your spaceship on landing is a profoundly bad day.

A preferred site should enable expeditions to conduct operations. That means the ability to do science at locations designated by NASA as Regions of Interest and to have access to local resources that maintain human life on the Red Planet. More to this last point, NASA stipulated this must-have: Any exploration zone should allow tapping into at least 100 tons of water for living-off-the-land purposes over a 15-year stretch of time.

Staying alive

NASA astronaut Stanley Love is front and center when it comes to his insights about future Red Planet travel. Point of fact, explains the space traveler: Astronauts don't care about *where* they go on Mars.

"But we do care about safety and operability," Love explains. "We care deeply about whether the site is going to kill us and whether we can perform the assigned work." The astronaut underscores a practicality: Landings on Mars "are very, very high-risk endeavors," he suggests, and once a crew is firmly footed on Mars, "just keeping alive takes a lot of time and effort."

Interestingly enough, Love adds, the need-to-have features of a Mars landing site for humans are the same as those required by the robots on Mars. There is one added concern, he notes as a sad fact, and that is "cooties." Space suits leak, habitats leak. "Out-gassing" on Mars from humans means bacteria and viruses percolating out onto the planet. And quid pro quo, Love says, just as Mars walkers cannot prevent leaking biota out onto Mars, there's no preventing Martian stuff from eking back into the home base habitat.

That being the case, there's a decision facing the humans-on-Mars exploration community, Love says. Should astronauts visit sites viewed as good candidates for exploring life on Mars, if it ever existed, or should crewed missions prioritize "planetary protection," taking pains not to introduce Earth life into places that might be the most likely for life to thrive?

Bottom line for Love: "It's a tough choice we have to make."

Where to land

MEANWHILE, SCIENTISTS KEEP EXPLORING MARS from afar to decide the target point for the first human landings. Rich Zurek, Mars Reconnaissance Orbiter project scientist at NASA's Jet Propulsion Laboratory in Pasadena, California, has been working with a team of experts to develop a specialized 2020 Mars orbiter loaded with an array of instruments, specifically powerful radar gear to identify underground reservoirs of ice. With those data in hand, the whereabouts and suitability of a variety of Mars sites for a human outpost can be advanced.

Landing site investigators envision robotic craft dispatched as early arrivals to the chosen site. Their duty is to ensure that the welcome mat is fully deployed for incoming astronauts, perhaps putting in place sections of early housing. Think of it as a robotic version of the Motel 6 mantra: Robots will get the site ready, and then "we'll leave the light on for you."

Spread out over several Mars missions, larger pieces of an outpost can be delivered. This steady cadence of arriving cargo ensures an expansive infrastructure to support crews. The pattern continues even as spacecraft bring human cargo as well. Each mission carries more supplies, and each crew rotation on Mars adds to the work of previous crews, resulting in a dwindling need in the long run for supplies shipped from Earth.

A planet ripe for processing

WHEN THE FIRST BOOT PRINT on Mars happens, the thought of gaining firm footing on that isolated world calls for living off the land. We already know that Mars is a plentiful planet, one that has exploitable resources to support any future expeditions.

But humans able to hunker down and live off the unearthly scenery that is Mars is easier said than done. What is required is adopting a theme dubbed as in situ resource utilization, or ISRU. That is space-speak for an on-the-spot technological path not only to endure Mars but also to thrive there. What's needed for humans to truly prolong a satisfactory, albeit bare-bones, lifestyle on the Red Planet?

ISRU on Mars is a work in progress, explains Robert Mueller, a senior technologist within the Science and Technology Projects Division at the Kennedy Space Center in Florida. First, water and atmospheric carbon dioxide on Mars are anticipated to be the most valuable resources for human missions, he says. Those resources can yield propellants for shuttling crews back to Earth via a Mars ascent vehicle.

Second, Mars-rich assets are ripe for the picking and processing; made-on-Mars products can be useful for life support, growing crops, even radiation protection.

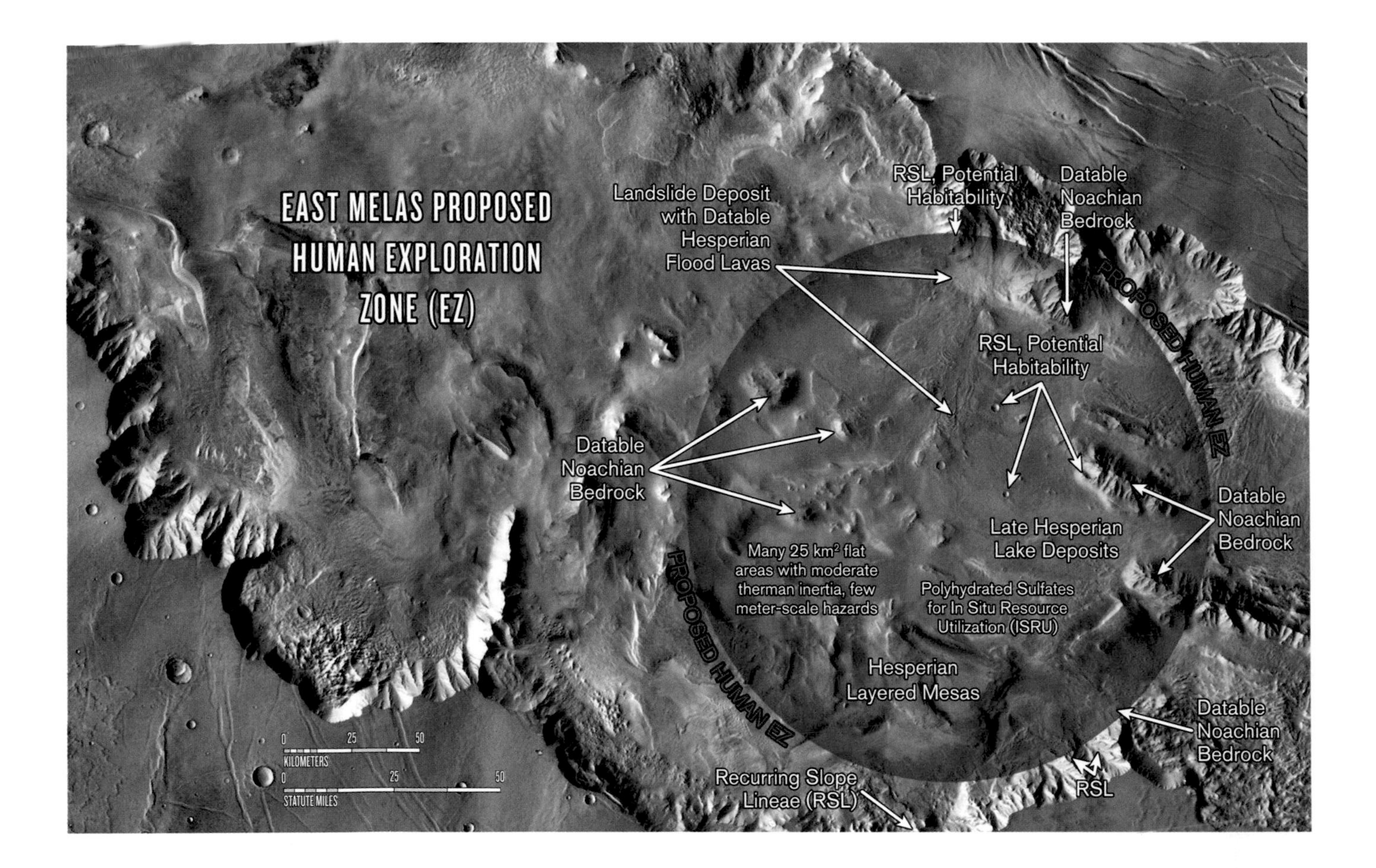

That said, the ultimate selection of a Mars habitat spot depends, in very large measure, on determining what resources are already there for the picking to support human life.

Mueller highlights the view that tapping local resources is vital to being truly Earth independent in space. There's a lot of work to do before this idea can become certainty. What are those resources, he asks, and are those resources economically and physically attainable?

Like-minded on the ISRU issue is Angel Abbud-Madrid, director of the Center for Space Resources at the Colorado School of Mines in Golden, Colorado. Martian resources are now considered the enabling element for a sustainable human exploration campaign, he observes, but adds one caveat.

Whereas scientific knowledge of the whereabouts of Mars resources is important, accessing them means devising the best methods to extract them, Abbud-Madrid suggests. Equipment is necessary, for example, to make propellants and radiation shielding, to keep thermal systems cool, to produce food, and to find potable water suitable and ample enough for human needs.

One proposed exploration zone for a human landing on Mars. Sites such as this contain regions of interest chosen for their scientific significance and resource potential, making them suitable for longer stays on the Red Planet.

Of that list, water is the most precious of them all, Abbud-Madrid says. "Such large quantities of that vital liquid are needed for an ongoing operation of a human outpost. Carrying that resource from Earth is not a viable option."

Water, water everywhere

THERE IS DEFINITELY GOOD NEWS concerning water on Mars. Several sources of water deposits on Mars have been recognized over the years, two of which appear reasonable sources of the essential water amounts: subsurface ice/permafrost or water bound to rocks and finely grained soil in the form of hydrated minerals, such as clays or gypsum.

Yet another prospect for water may be the newly discovered seasonal flows of water called recurring slope lineae, tagged simply as RSL. These RSLs have been branded as signs of intermittent flowing liquid water, although briny in makeup. Still to be ascertained is whether RSLs yield water usable for human consumption.

A number of locales on Mars appear ready to serve up feedstocks of water. Yet the amount of energy requisite to extract water—be it from seasonal flows of RSL water, sheet ice, glacier ice, or hydrated minerals/adsorbed water—has not been determined or studied.

Abbud-Madrid observes that determining the source type, location, depth, distribution, and purity of potential water deposits on Mars is basic homework. Beyond that, engineers must develop the most resourceful techniques for excavating, beneficiating, extracting, and purifying water on the Red Planet—and judge the energy needed to execute those operations.

DISASTERS

Thrown off course | An off-kilter burn out of Earth orbit, magnetic storms, or a coronal mass ejection from the sun, a stray asteroid: Unexpected events could throw a Mars-bound vehicle so far off course that it would be difficult to reorient.

What could go wrong?

Staying on Mars

GETTING THERE IS THE FIRST STEP. Staying there is the next. It is the outlook of those blueprinting a "staying power" strategy of humans on Mars that use of on-the-spot Martian resources is crucial. A costly one-time trip heaving all required supplies from Earth makes exploration of the Red Planet unsustainable.

Scripting the first human trek to Mars harkens back to the heyday of the Apollo program and vaulting astronauts to the moon. But there are also differences, suggests Jim Head of Brown University in Providence, Rhode Island. He is no stranger to humans waving goodbye to Earth and heading for new destinations; he worked on

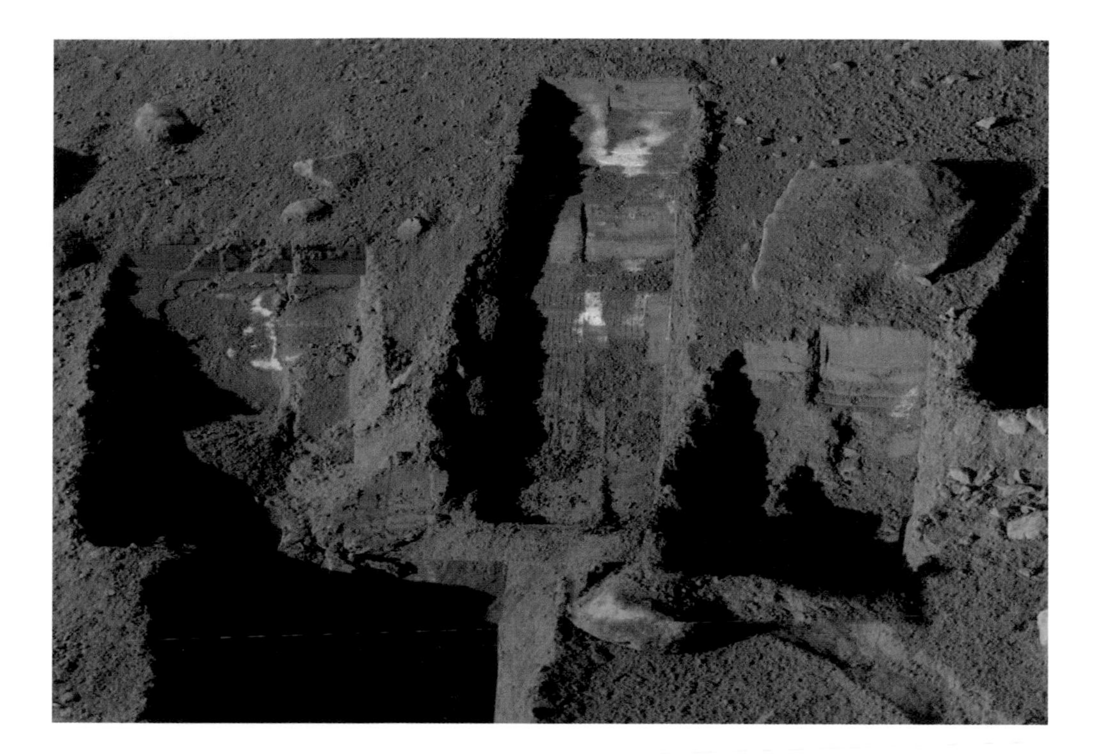

After using its robotic arm to dig a ditch named Snow White, NASA's Phoenix Mars lander took this shadow-enhanced false color image in October 2008, showing morning frost and subsurface ice within the trench.

the Apollo program by evaluating potential landing sites and training crews for their lunar tasks.

For Apollo, Head recalls, there was a fast-paced immediacy fueled by the space race between the United States and the Soviet Union. "For Mars, we've got time," he says, and the good news is that this go-round permits creation of long-term science and engineering synergism—the technological right stuff that made Apollo lunar exploration achievable and productive.

"We need to live first off of Mars," Head concludes, "so we need sustenance there by using local resources that make real cutting the umbilical cord between Earth and Mars . . . because it's going to be cut someday."

In the opening years of human treks to Mars, a semipermanent base can be built. As ground-truthing expeditions judge the reservoirs of available water and other resources, pioneering on the Red Planet is anticipated to result in an Earth-independent, extended-stay capability for humans there. To guarantee that outcome, efforts have begun both on and off Earth to gauge the physical and mental stresses of making Mars a home away from home.

No doubt Mars itself will toss woes and worries at initial voyagers. The human psyche throws in its own complications. What are the biomedical and sociological bugaboos and barriers that need to be defeated before protracted stays on that intimidating world are a given? ■

TEST LAUNCH

A Delta IV Heavy rocket departs Cape Canaveral, Florida, on December 5, 2014, lofting an uncrewed Orion spacecraft on Exploration Flight Test-1. The Orion orbited the Earth twice, reaching speeds of 20,000 miles an hour and traveling through belts of intense radiation. Then, after reentering Earth's atmosphere, it parachuted into an ocean landing area for recovery.

MARS, HERE WE COME

An Atlas V rocket clears the tower at Cape Canaveral, Florida. The November 2011 launch hurled the Mars Science Laboratory mission's SUV-size Curiosity rover toward the Red Planet. In August 2012 the world watched as Curiosity successfully landed on Mars.

PRACTICE MAKES PERFECT

NASA astronaut Scott Kelly trains inside a Soyuz simulator at the Gagarin Cosmonaut Training Center in Russia. Russian cosmonaut Mikhail Kornienko and Kelly together flew a nearly year-long mission on the International Space Station, both returning to Earth in March 2016.

ОТКР.

HEROES | JANINE CUEVAS

Lead Material Requirements Planner, Aerojet Rocketdyne, NASA's John C. Stennis Space Center

Safely placing cargo and humans on Mars is a huge technological task, says Janine Cuevas. She should know; the NASA employee has been working on America's launch vehicles for nearly 30 years.

Cuevas is involved with a major NASA undertaking to hurl habitats and crews to Mars, represented by construction of the space launch system (SLS). This colossal booster is designed to carry crews of up to four astronauts in the space agency's Orion spacecraft and send them to multiple, deep space destinations, particularly Mars. The SLS will be the first exploration-class launch vehicle built since the Saturn V rocket that lobbed astronauts to the moon in the late 1960s and early 1970s.

"We currently have a fleet of 16 flight-worthy liquid propulsion rocket engines from the shuttle program," she says, "and being able to upgrade those engines and adapt them to power the SLS core stage will be an enormous achievement." Cuevas is now overseeing efforts to reconfigure the earlier space shuttle main engine for use in the SLS initiative. "I am responsible for making sure the correct hardware configuration is available in the right place, at the right time," she says. Those repurposed and modified Aerojet Rocketdyne RS-25 engines will support initial missions of NASA's SLS and have the ability to lift 77 tons. That's more than double any operational rocket today.

During the space shuttle era, Cuevas worked as a lead mechanical technician on the space shuttle main engines. Assembling and testing those powerhouses was critical, she explains, knowing full well that astronauts' lives depended on the product that was delivered for flight. "In my role as a technician, I never realized how many details were involved in getting hardware delivered to the assembly floor," Cuevas points out. "I just knew that we needed it to be there in time to support the proper assembly sequence."

There is a palpable countdown under way to ready RS-25 engines for the SLS. In January 2015, the first "hot-fire" test was performed: engine fired and run without liftoff. Other roaring trials have followed, also carried out at the Stennis Space Center, to accumulate test data. On the books for 2018 is the first SLS exploration mission test flight; an uncrewed Orion spacecraft will sit atop the megabooster, blasting off from a newly refurbished site at the Kennedy Space Center in Florida. In future years, as the SLS evolves, it is slated to provide a lift capability of 143 tons.

Today, regardless of one's role on the SLS program, says Cuevas, "it must be considered a high degree of difficulty . . . because there is no room for error on this mission."

An RS-25 engine—a class of rocket motor that successfully powered the space shuttle—roars to life. The engine is now being modified for use in the NASA space launch system that will ultimately send cargo and crew to Mars.

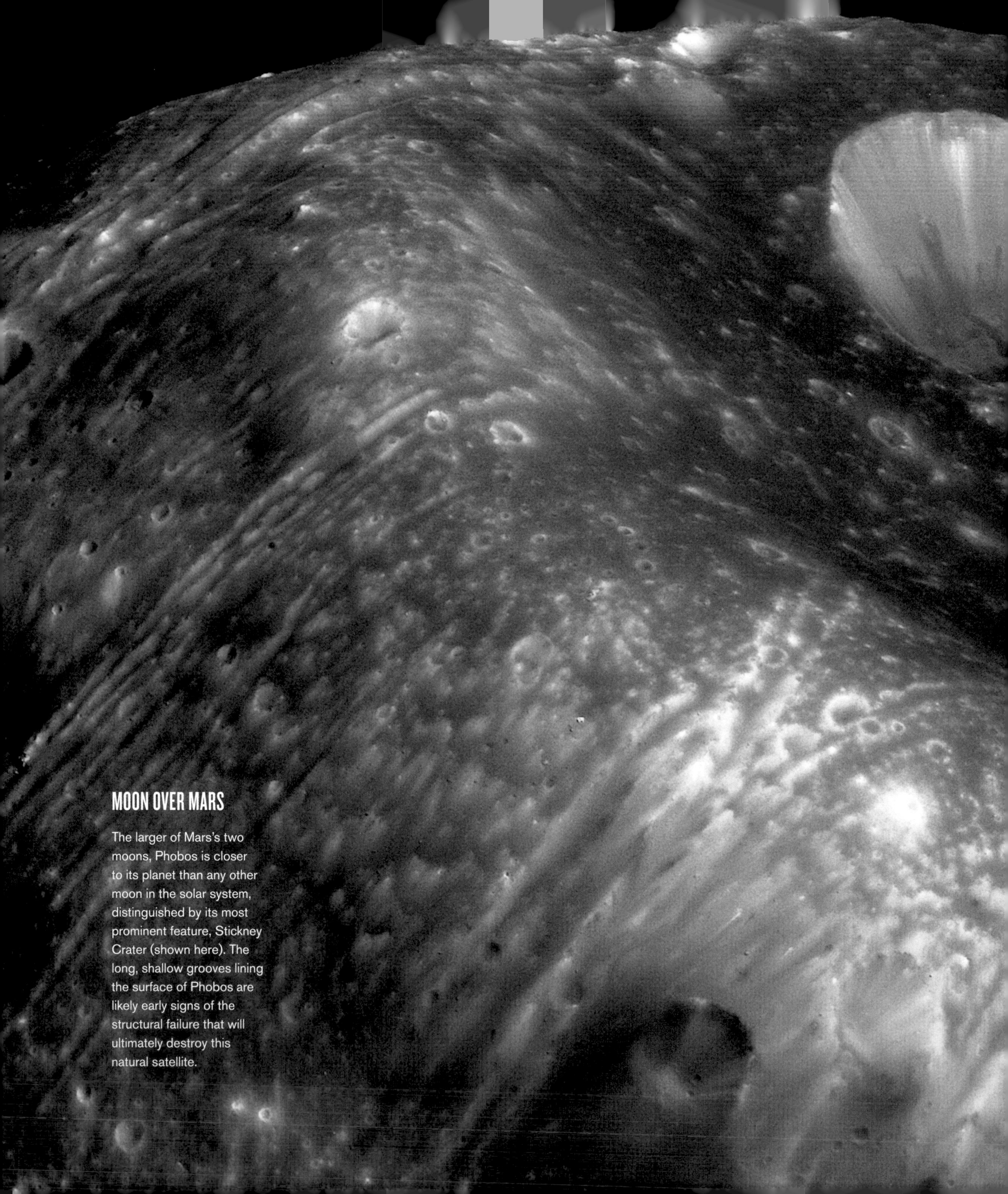

MOON OVER MARS

The larger of Mars's two moons, Phobos is closer to its planet than any other moon in the solar system, distinguished by its most prominent feature, Stickney Crater (shown here). The long, shallow grooves lining the surface of Phobos are likely early signs of the structural failure that will ultimately destroy this natural satellite.

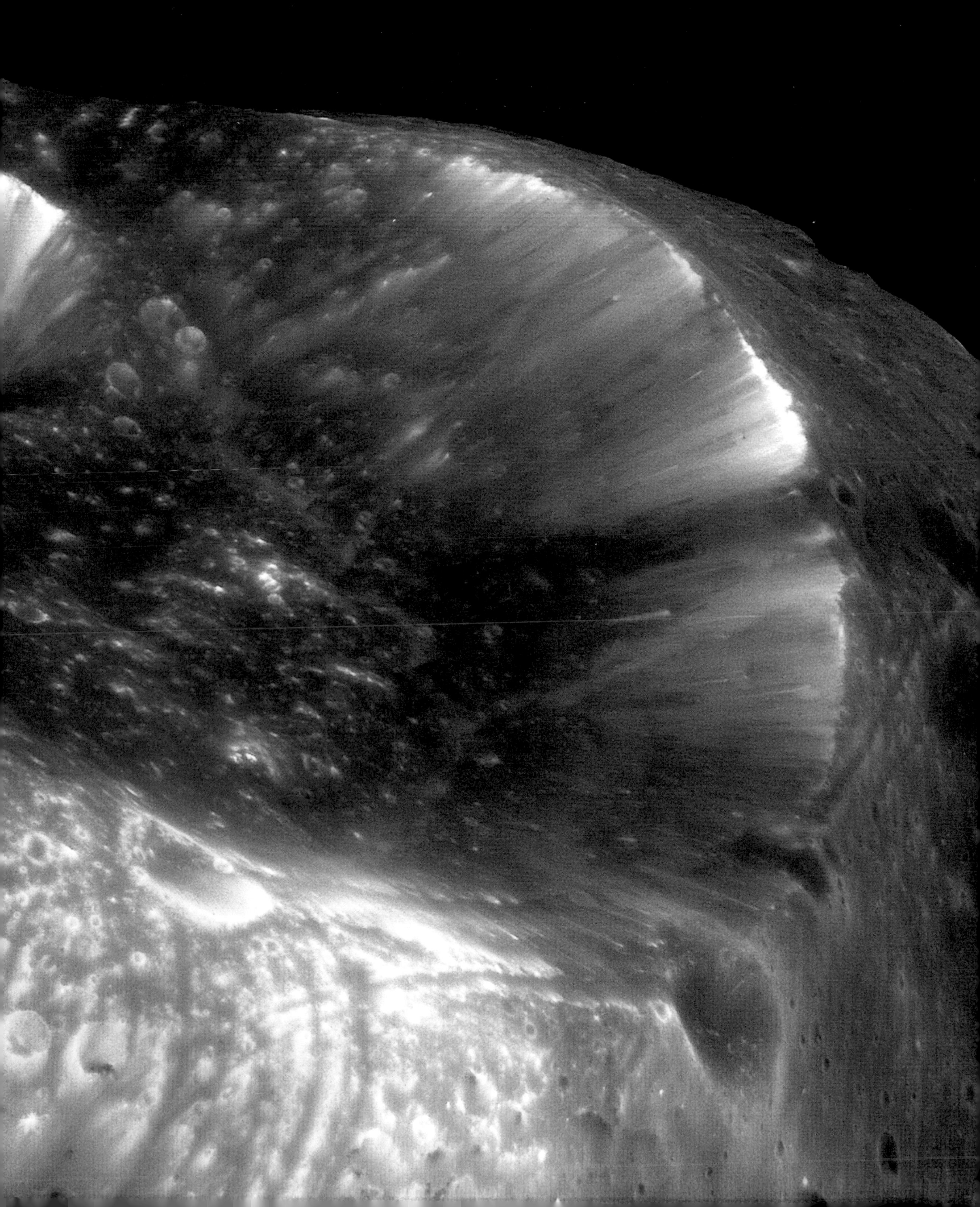

PUTTING ON THE BRAKES

Testing is under way to investigate methods for landing expeditionary crews and heavy habitats on Mars. Hypersonic inflatable technology (right) makes use of flexible materials that protect a spacecraft from the intense heat experienced during entry into the atmosphere. Another landing technology under study—the low-density supersonic decelerator (below)—is essentially a supersonic parachute.

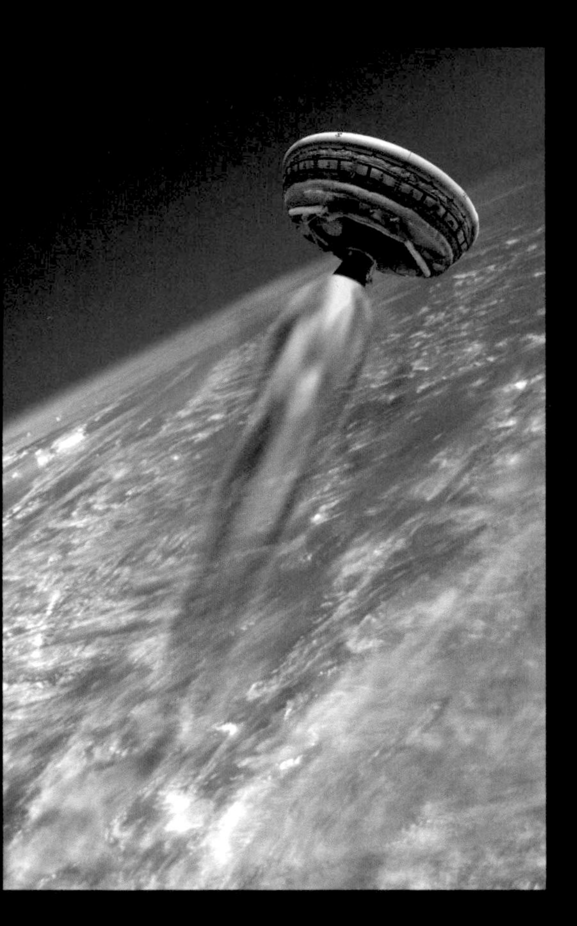

BEFORE OUR VERY EYES

Mars is a planet pocked with strike marks from meteorites or other stray space material. Recently the Mars reconnaissance orbiter showed a new impact crater, formed between July 2010 and May 2012. The crater itself is nearly 100 feet across, and debris sprayed as far as nine miles around it.

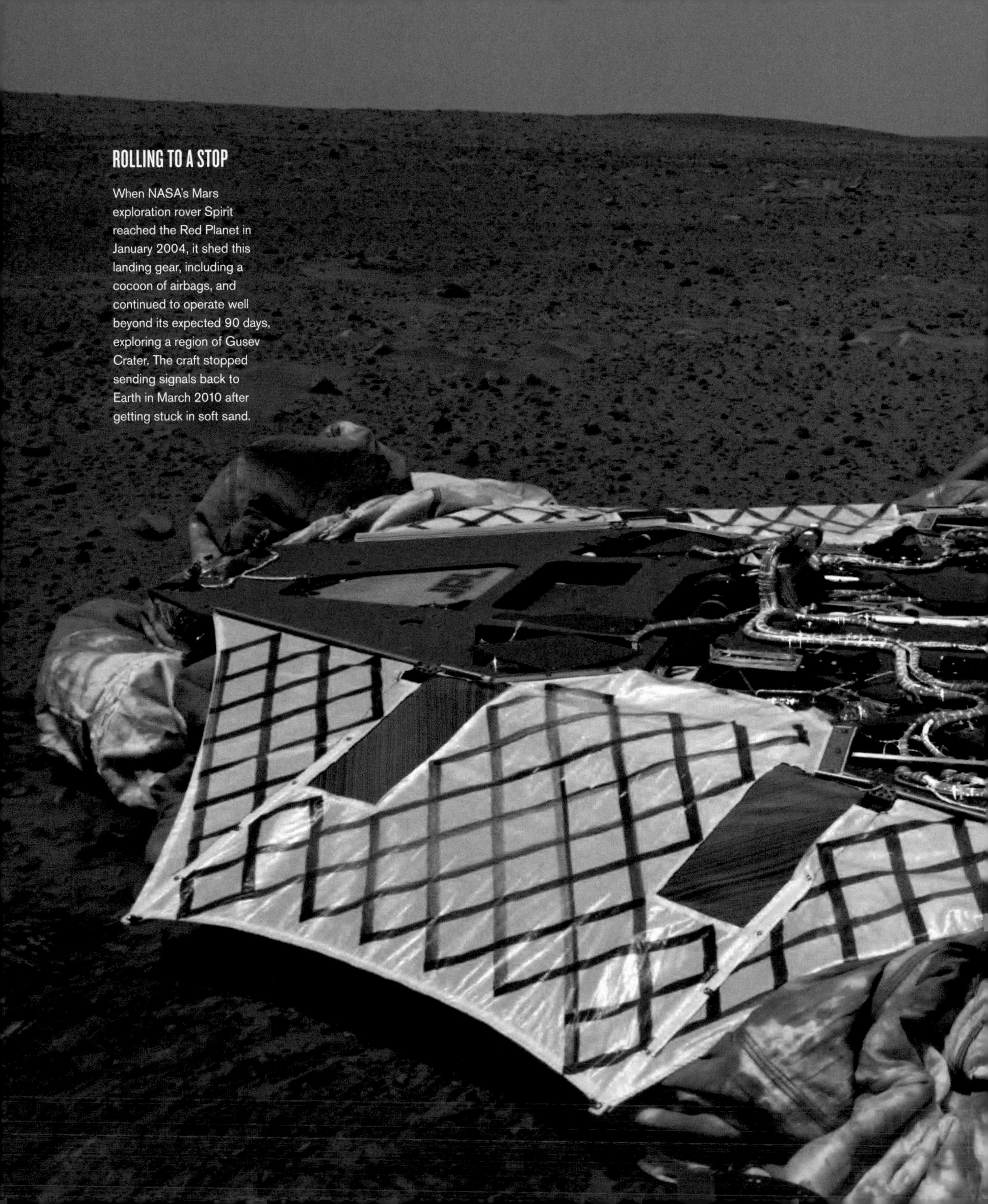

ROLLING TO A STOP

When NASA's Mars exploration rover Spirit reached the Red Planet in January 2004, it shed this landing gear, including a cocoon of airbags, and continued to operate well beyond its expected 90 days, exploring a region of Gusev Crater. The craft stopped sending signals back to Earth in March 2010 after getting stuck in soft sand.

KICKING UP DUST

For the 2012 Mars landing of the one-ton Curiosity, engineers devised a novel contraption called the sky crane. After an entry and parachute phase, the rocket-powered descent stage hovered above the planet (left), then lowered the rover on a bridle (below) and delivered it to the surface of the Red Planet.

SMILE FOR THE CAMERA

NASA's Curiosity Mars rover took dozens of images of itself that were combined to create this history-making selfie, posted to the mission's Facebook page with the caption "Hello, Gorgeous!" The nuclear-powered rover has been busily reconnoitering the planet since August 2012.

HEROES | ROB MANNING

Mars Engineering Manager, Mars Program Office, NASA's Jet Propulsion Laboratory

For the engineers who dreamed up how to land NASA's Curiosity rover on Mars, a degree of edginess still remains when they recall the robot's final minutes before making a safe touchdown on the Red Planet in August 2012.

"None of this stuff is a slam dunk," says Rob Manning, an entry, descent, and landing specialist for Mars missions at California's Jet Propulsion Laboratory. His engineering know-how has been used on nearly every U.S. Mars mission of the past 20 years. Shipping huge payloads to Mars—such as habitation modules and piloted craft—is no simple outing, he emphasizes, and the supersonic parachute used when Curiosity met Mars is not part of the larger equation. It would be far too big to unfurl effectively and dependably. Instead, work is in progress to be able to slam on the brakes at Mars with a hypersonic inflatable aerodynamic decelerator device. Supersonic retropropulsion—roaring rocket engines—would then take over for the final soft landing.

To realize boots on Mars, Manning takes a first-things-first stance. He sees the pioneering mission as a "flags and footprints" undertaking. People on that first flight, he says, will know that they are taking a huge gamble. "If you're climbing a mountain, you've got to do the first steps first," Manning says. "You can't do the top of the mountain first. You have to work your way up," with successive expeditions adding to that foundational flight.

What happens if that premier human landing goes awry? "Failure puts the brakes on our progress. But it is also key to our progress," he answers. "If cost was no object, I think that the very first landing event on Mars would be uncrewed and utilize the exact system we'd use for the first human mission. If it did fail, we would likely prevent the political backlash. You would still have to stand down and rethink everything. At least we would know what happened and the politics won't be so dire."

Advice for what gives a person the "right stuff" in rocketry headed for Mars? A deep curiosity, according to Manning, as well as a willingness to learn and try, and fearlessness. "We've got to be able to do not less, but more. The trick is to keep things simple, use the technology to your advantage, and keep the price under control." Redundancy is key. "We have to think through all those 'what if' scenarios that really allow you to win in the long run. It's like playing poker. To consistently win at poker, in our game, you have to hide the aces in every sleeve you've got."

Another self-portrait of NASA's Curiosity Mars rover, this one of its left wheels taken by MAHLI, the rover's Mars hand lens imager, designed for detail work. The rover's treads imprint "JPL" (for Jet Propulsion Laboratory) in Morse code into the Martian dust.

READYING THE ROBOT

About the size of a small SUV, NASA's Curiosity rover has six-wheel drive and the ability to turn in place a full 360 degrees, as well as the agility to climb steep hills. The rover includes chemical lab facilities designed to investigate whether its landing site within Gale Crater ever offered conditions favorable for microbial life.

TALES DUNES TELL

Orbiting cameras have been monitoring dunes on Mars as they shift and shimmer, telling us more about surface topography and wind behavior on the planet. More than 60 sites have been captured in images such as this (shot through a blue filter). Comparisons through time show that sand dunes on Mars can move more than a meter in a single sol, or Martian year.

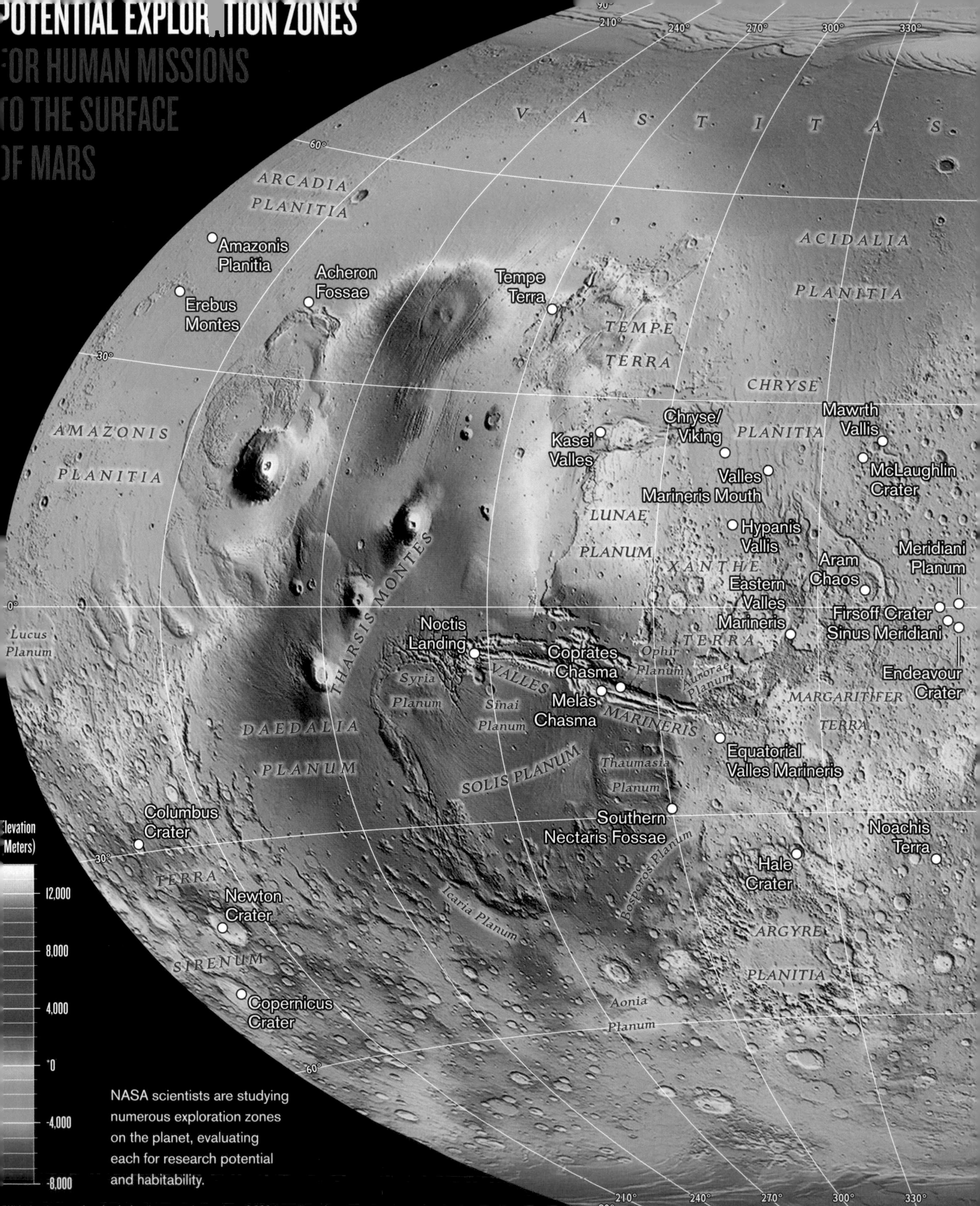

NASA scientists are studying numerous exploration zones on the planet, evaluating each for research potential and habitability.

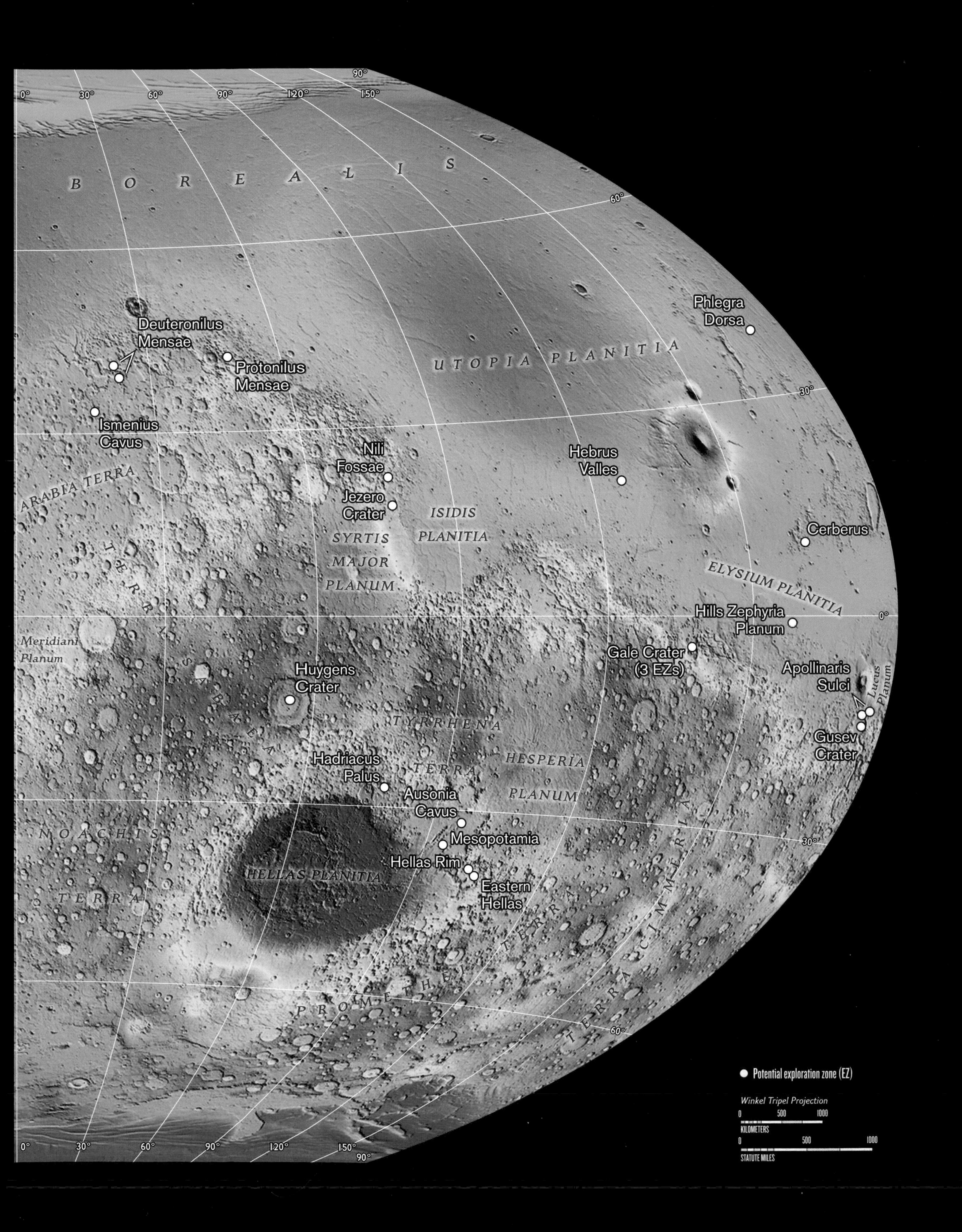
B O R E A L I S
Deuteronilus Mensae
Protonilus Mensae
Ismenius Cavus
Phlegra Dorsa
UTOPIA PLANITIA
Nili Fossae
Jezero Crater
Hebrus Valles
ARABIA TERRA
ISIDIS PLANITIA
SYRTIS MAJOR PLANUM
Cerberus
ELYSIUM PLANITIA
Hills Zephyria Planum
Gale Crater (3 EZs)
Meridiani Planum
TERRA SABAEA
Huygens Crater
Apollinaris Sulci
Lucus Planum
Gusev Crater
TYRRHENA TERRA
HESPERIA PLANUM
Hadriacus Palus
Ausonia Cavus
Mesopotamia
Hellas Rim
Eastern Hellas
NOACHIS TERRA
HELLAS PLANITIA
TERRA CIMMERIA
PROMETHEI TERRA
Potential exploration zone (EZ)
Winkel Tripel Projection
KILOMETERS
STATUTE MILES

GRAND VIEW OF HISTORY

Valles Marineris, dubbed the Grand Canyon of Mars, averages more than 100 miles wide, its floor layered with rocks and debris within which we may one day read Mars's geological history. This image is a pastiche of many frames captured by Mars Odyssey, orbiting since 2001.

Physical challenges aside, mental and emotional stresses abound as humans make their home on a new planet.

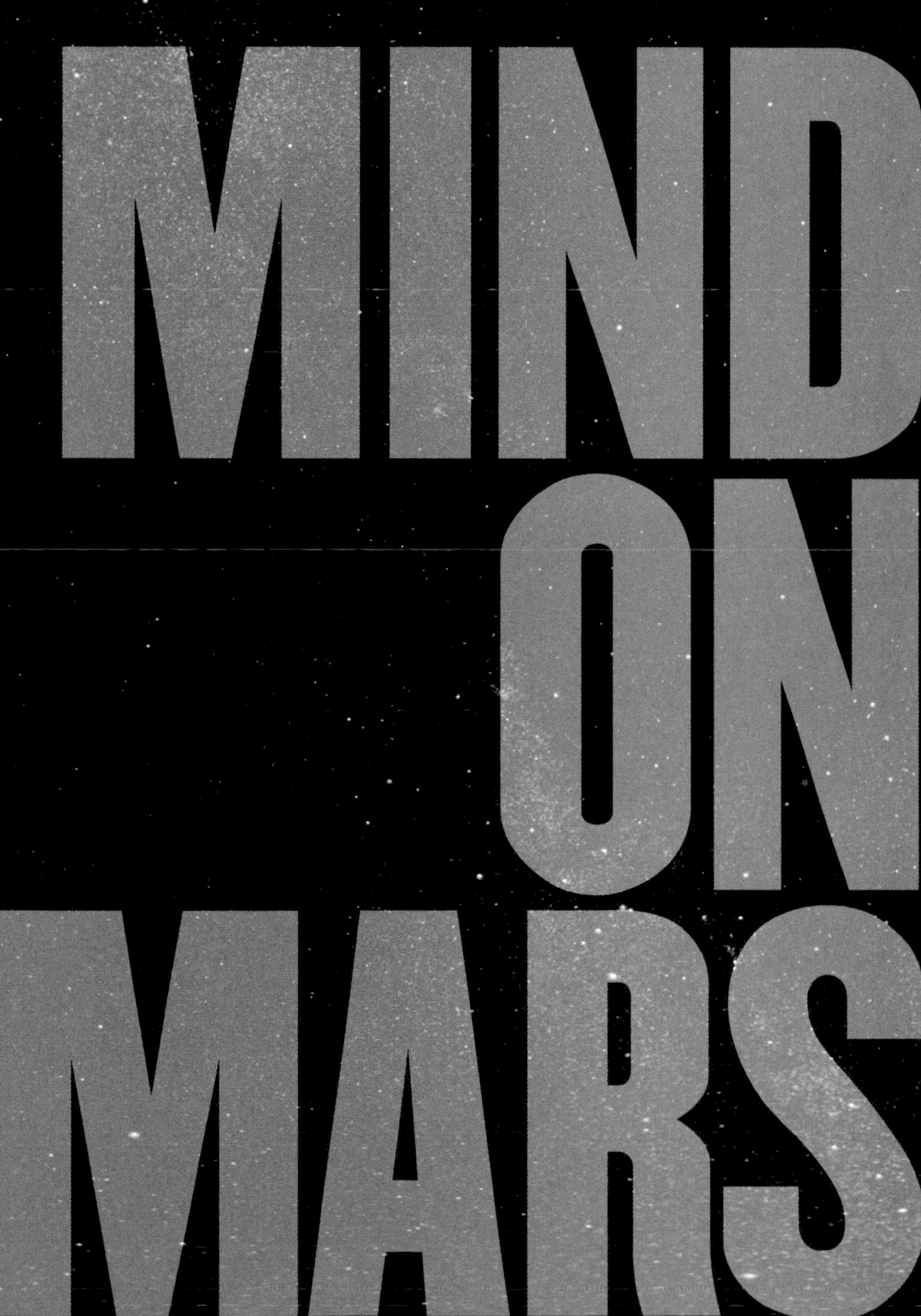
MIND
ON
MARS

NASA astronaut Karen Nyberg, flight engineer on the International Space Station's 2013 Expedition 36/37, gazes homeward toward Earth from the station's cupola windows.

2

Mind on Mars

HE VOYAGE OF A CREW TO MARS is a protracted, perilous one. It is an interplanetary adaptation of the loneliness of a long-distance runner. Confronting and enduring the emotional and mental stresses and strains of just getting to the Red Planet is rough enough, putting aside the psychological tensions of chalking up any lasting stay on Mars.

Who should go to Mars, and are they fit for flight? What's in the inner-personal travel kit that constitutes the most righteous of the right stuff to make the sojourn? There are early tip-offs about how humans cope with and living in extreme environments.

In many ways, individuals are already training for the long passage to Mars. The International Space Station has become psychosocial central in evaluating how lengthy stints in space can affect space travelers. A historical fact: One American Skylab crew in the 1970s "went on strike" because of too many demands placed on them by mission controllers. Flying aboard the final Skylab mission that lasted 84 days, the crew bemoaned that they were overscheduled, harried, and hustled, letting ground control know in no uncertain terms of their concerns before instituting a daylong work strike.

Zinnias grown in a controlled environment chamber at NASA's Kennedy Space Center in Florida—harvested at the same time Scott Kelly picked zinnias he grew aboard the International Space Station—provide information on how to grow crops for food on Mars missions.

More recently a travelogue of teachable moments was logged in by two ISS residents, U.S. astronaut Scott Kelly and Russian cosmonaut Mikhail Kornienko. The two made a groundbreaking endurance mission spending nearly a year on the orbiting outpost. How they dealt with and offset the sense of isolation has provided clues to carrying out future Mars missions.

For NASA, identical twin astronauts Scott and Mark Kelly took part in an investigation cleanly labeled "The Twins Study." The aim of the off-Earth inquiry was to observe space-flight-induced effects and changes that might occur in astronaut Scott

EPISODE 2

GROUNDED

The mission has only just begun, but it is already at risk. One *Daedalus* crew member is badly hurt, and in order to help him, the rest of the team must reach their base camp, dozens of kilometers away, where previous unmanned flights have deposited the resources they will need to survive. Guided by flight control back on Earth, the mission's second in command must lead the crew, forced to push their equipment and their bodies to the limit as they cross deadly Martian landscape.

Kelly compared to his Earth-bound twin, Mark Kelly. This novel appraisal of two individuals who have the same genetics but have lived in different environments for one year was set up as a multifaceted national research project coordinated with expertise resident in universities, corporations, and government laboratories.

The research included such life science issues as: How does the human immune system change in space? Does space radiation prematurely age space travelers? What's the impact of microgravity on human digestion? Why do astronauts report changes in vision? And what's the phenomenon called "space fog"—a lack of attentiveness and the slowing of mental gears reported in the past by some astronauts in Earth orbit?

Guesstimating space radiation risks for any Mars-bound crew remains a front-and-center concern. Fatal cancer risk is a worry for astronauts traveling beyond Earth. In their travels hugging the planet in low Earth orbit, as in the International Space Station, astronauts are partially protected by the Earth's magnetic field and the solid body of the planet itself. But departing for Mars is another matter: Astronauts become naked to nature. Some research points to radiation risks to the central nervous system, even an acceleration of Alzheimer's disease.

Imagine going to and returning from Mars and not remembering the escapade. That would seemingly be a nonstarter for many. There's a long list of medical questions that need dedicated study prior to sending humans to the Red Planet.

Close-quarters seclusion

WHILE THE INTERNATIONAL SPACE STATION is seen as a starting point for undertaking 21st-century deep space travel, a number of analog sites on Earth are also prompting advisory notes before humans congregate in the harsh climes of Mars. From the High Arctic to the South Pole and submarines, each of these remote locations offers characteristics that mimic Mars travel and help hasten its day. Similarly, isolation chamber research, particularly in Russia, has put individuals through mock trips to the Red Planet.

Perhaps one of the more inventive psychosocial isolation experiments was Mars500, a cooperative project between the European Space Agency and the Russian Institute for Biomedical Problems. It was conducted in stages between 2007 and 2011, with the isolation facility for crews situated in a special building on the Russian Institute site in Moscow. Mars500 concluded with a record-setting 520-day simulated mission to Mars. The crew members, all of them men, were three Russian, one French, one Italian, and one Chinese citizen. The facility consisted of the isolation facility itself, as well as the operations room, technical facilities, and offices. The isolation structure had four hermetically sealed, interconnected habitat modules totaling nearly 20,000 cubic feet. The creative project simulated an Earth-Mars shuttle spacecraft, and an ascent-descent craft. The crew used an external module to stroll about on a simulated Martian surface.

Results from Mars500 underscored that the crew experienced a pleasant and harmonious time with each other. Still, given their isolation, they sorely missed family and friends, as well as contact with unfamiliar faces and other viewpoints. From an engineering side, biofilms—thin, tough layers of bacteria adhering to one another—formed inside the mock spacecraft and on elements in the life support system. These could present risk of infection of long-term space travelers, perhaps even lead to malfunction of instruments, reported investigators from the German Aerospace Center (Deutsches Zentrum für Luft- und Raumfahrt).

Foothold on the future

THE NOW ORBITING INTERNATIONAL Space Station is acknowledged as the most complex scientific and technological endeavor ever undertaken to date. Some tag it as a "weightless wonderland." Since on-orbit assembly of the complex started in 1998, 16 nations have contributed to realizing and using this outpost in space. It now has more livable room than a six-bedroom house, spread across 14 pressurized modules or components. Its internal volume is about the same as a Boeing 747 jumbo jetliner, its overall length and width about the size of an American football field.

"What are the important steps in the evolution of life? Obviously the advent of single-celled life. Differentiation to plants and animals. Life going from the oceans to land. Mammals. Consciousness. And . . . also on that scale should fit life becoming interplanetary."

—Elon Musk, Founder and CEO of SpaceX

It has three laboratories—the U.S. Destiny module, European Columbus module, and Japan's Kibo lab—and three connecting nodes, Unity, Harmony, and Tranquility. On the Russian side are two docking compartments (Pirs and Rassvet), the Zarya FBG, and the Zvezda service module.

Station modules stuffed with equipment have made feasible active studies by space station crews on human health in microgravity, biological processes and biotechnology, Earth observation, space science, and physical sciences. Years of construction and early checkout behind it, the International Space Station is arguably a foothold on the future; coming and going crews are testing technologies, systems, and materials that provide the know-how for long-duration missions.

Bitter truths

RESEARCH STATIONS IN ANTARCTICA are ideal places to study how humans adapt to living in far-flung, isolated locales. Studies there are helping to appreciate the effects of space travel and life inside habitats on Mars, as astronauts embark on long periods in space and are cut off from our world, starved of sunlight, and surviving in small communities.

A case in point is the British Antarctic Halley Research Station that hosts studies with the European Space Agency (ESA) to rate human adaptation to space travel. Depending on the time of year, the facility houses between 13 and 52 scientists and support staff. The bitter truth is that the station in winter drops to temperatures as low as minus 58°F and sees more than four months of darkness.

One project that ran at Halley for months had team members recording themselves in a video diary. Those diaries are analyzed with a computer algorithm through parameters such as pitch and word choice. Researchers expect that this technique will provide a new window to objectively monitor a person's psychological state and how well he or she has adapted to the stresses of prolonged spaceflight.

ESA participates in Concordia, the Italian-French base in Antarctica, using the base as an extraterrestrial look-alike for habitats on other worlds. Indeed, the ice island that is Concordia is nicknamed "White Mars." This compound is far removed from civilization, like no other place on Earth. Getting to Concordia station can take up to 12 days by plane, ship, and caravans on skis. ESA researchers underscore the fact that the nearest human beings are stationed over 370 miles away at the Russian Vostok base, making Concordia more secluded than the International Space Station is from Earth.

Studies of the effects of isolation in multicultural crews at Concordia are providing ESA useful facts for a mission to Mars. The base serves as a laboratory for medical monitoring and testing life support technologies. Aside from hardware properties such as weight, strength, and resistance, astronauts need surroundings free of harmful bacteria and mold. ESA investigators are assessing which materials are best suited to build spacecraft and are testing a variety of antimicrobial samples in Concordia.

Make-believe Mars

Devon Island is the world's largest uninhabited island. The remote site of NASA's Haughton Mars Project (HMP), it is a fitting analog for Mars: a polar desert within the Canadian High Arctic.

"If you were to describe Devon Island, it is cold, dry, rocky and barren, cut by canyons, valleys and gullies, packed with ground ice, and scarred by impact. Now that's also how you would describe Mars," says Pascal Lee, the project's mission director. "No place on Earth is exactly like Mars. But places like Devon Island present specific similarities that help you get one step closer."

Analogs of Mars here on Earth serve several purposes, says Lee. "They help you *learn, test, train,* and *educate.* Devon Island has already been helping us *learn* about Mars and how to explore it, *test* new exploration technologies and strategies, and *educate* students and the general public."

Other analog activity is under way by the Mars Society. This private group, based in Colorado, has initiated the Mars Analog Research Station project, including use of two Mars base-like habitats, one planted on Devon Island in the Canadian Arctic and the other in the American Southwest. In these Mars-like environments, the Mars Society regularly performs extensive long-duration field exploration campaigns. Those operations mimic the lifestyle and common constraints that explorers will meet on Mars.

A great deal has been uncovered from their Mars mission simulations in the desert and in the Arctic, says Robert Zubrin, president of the Mars Society and cowriter with Richard Wagner of the influential 1996 nonfiction book *The Case for Mars: The Plan to Settle the Red Planet and Why We Must.* The book and a 2011 updated version is a technological tour de force that plots out "Mars Direct," a how-to proposal that reduces in cost and complexity a humans-to-Mars effort, one that uses Mars resources and methods to live off the land.

Over the last several years, Zubrin and researchers at the Mars Analog Research Station have pushed forward ways to make the Red Planet ripe for human investigation. "We have learned that the mission needs to be led from the front . . . so the team on Earth needs to understand that its role is mission support, not mission control," he emphasizes.

Mars-on-Earth re-creations suggest that the crew will need to be selected as a team, reports Zubrin. There have been many instances where an individual was a strong member in one crew but a problem in another because the chemistry between people was wrong. "So when the time comes to choose a Mars crew, we should let the shrinks take their best guesses at forming three candidate crews." That done, each team would be sent to a Mars simulation station in the Arctic or the desert and tasked with conducting a sustained program of field exploration for at least six months under Mars mission constraints, he suggests.

"We'll see which crew does that the best. That's the team we will want to send to Mars," Zubrin says.

Other Mars Society findings are that small, nimble field mobility systems like all-terrain vehicles are much more useful for a Mars crew than large pressurized rovers. The problem uncovered is that when big land rovers get stuck, you cannot get them moving again. "As far as Mars field equipment is concerned, if you can't lift it, don't bring it," Zubrin relates. Simple, robust scientific equipment is much more useful than highly complex but vulnerable sophisticated instruments that are nominally more capable. The take-away message, he suggests, is that a prospector needs a pack mule, not a racehorse.

"We have learned that an electronic navigation capability is extremely important. It's easy to get lost in a desert wearing a space suit. We don't need GPS on Mars . . . but the crew should at least lay out a set of radio beacons as one of its early tasks," Zubrin points out. "We have been confronted by the fact that hiking around in a space suit doing field exploration is a very physical activity, and the amount of it that a crew can do will depend a lot on its physical condition."

That Mars Society judgment carries with it a strong proviso to Mars mission planners: Find a way to get to Mars that does not leave the travelers in microgravity, or weightless, through the lengthy journey. "We can certainly survive a trip to Mars in zero g, but that misses the point. We are going to Mars to explore, not to say we did it. That means we should go to Mars in an artificial gravity," Zubrin contends. NASA's space medicine program, which is currently centered almost entirely on zero-gravity health effects, "needs to be reoriented," he cautions, to pay more attention on the physical implications of space travel in a gravity situation.

Finally, with Mars analog studies backing him, Zubrin flags an extra result: "The people who say that on a long-term Mars mission 'the human psyche will be the weakest link in the chain' are wrong. Our crews have proven very adaptable, and I see no reason why NASA crews won't be as well. The spirits of those who want to do this are strong. Provided we choose the right team for a piloted Mars mission, the human crew will be the strongest link in the chain," he concludes.

Keeping them alive and sane

THE VIEW FROM THE SLOPES of Hawaii's Mauna Loa volcano at 8,200 feet above sea level is picturesque. And you're also closer to Mars. That setting is home base for

HI-SEAS, standing for Hawaii Space Exploration Analog and Simulation mission. Since 2012 this project has been funded by NASA's Human Research Program. Multiple universities are taking part in the endeavor too.

The cozy HI-SEAS habitat offers some 13,000 cubic feet of habitable space. It comes complete with a usable floor space of approximately 1,200 square feet. There are small sleeping quarters for a crew of six, along with a kitchen, laboratory, bathroom, simulated air lock, and work areas. A large solar array unit located south of the habitat powers the facility. A backup hydrogen fuel cell generator is nearby as well. Even on Mars, this domicile would have curb appeal.

Billed as the longest NASA-funded Mars simulation in history, the first one-year isolation mission for HI-SEAS places it in the company of a small group of analogs that are capable of operating very long duration missions (eight months and longer) in an out-of-the-way and confined environment. A team of approximately 40 volunteers from around the world serves as HI-SEAS mission support, interacting with the crew through imposed one-way 20-minute communications—a restriction devised to mimic the ambience of Mars life more closely. Exploration tasks by crew members include space-suited geological fieldwork saunters outside the habitat.

Primarily the project is examining crew composition and cohesion, gaining experiences that those on a planetary surface exploration mission will face. Studies target the psychological and psychosocial factors that help ensure highly effective teams for self-directing, long-duration space outings in the future.

"Basically, we're looking into how to keep them alive and sane . . . not wanting to kill each other in a long Mars mission," says Kim Binsted, the HI-SEAS project principal investigator and professor at the University of Hawaii at Manoa campus. It'll take a while to uncover results, but one preliminary item is not unexpected. "On long-duration missions there's always going to be conflict. It's really not avoidable," she counsels, be it from leadership squabbles to someone just stomping off in a huff over missing favorite food.

So how best can a Mars team have the ability to recover from conflicts, then come back and maintain a high level of performance? That's part of the HI-SEAS research agenda. "There's no going out for a beer," says Binsted. "There's no leaving each other alone for six months. You can't do avoidance . . . that's not an option."

Another problem is crew-ground disconnect, Binsted adds. It comes about partly due to time delay in communications. That factor ties into faraway crews' having autonomy. People on Mars will have more control over what they do on a day-to-day basis. That's

DISASTERS

Unknown terrain | Geological and meteorological catastrophes unknown on Earth could plague us on Mars: dust storms, ice-spewing volcanoes, earthquakes, landslides, lava tube collapse. We know very little about the forces on and below Mars's surface.

What could go wrong?

a far cry from the circumstances crew members see on the International Space Station. "Their daily schedule is defined down to the minute, and everything goes by ground control. On a Mars mission that just won't work . . . it won't happen," she stresses.

Binsted explains that HI-SEAS crews are testing equipment, evaluating protocols, even evaluating communications software concepts. No twiddling thumbs and being simply classed as guinea pigs, she notes, with NASA keen on acquiring early warnings on what can go wrong on a Mars mission. The space agency has oodles of risk categories. Some are colored green meaning under control; yellow might be a problem but it's a small chance difficulty or not high-impact trouble; a red risk is a showstopper and needs addressing.

"A subset of those red risks can be tackled with Mars analog studies. That's what we're trying to do . . . move those red risks over to the other columns," Binsted points out. Time is on their side at HI-SEAS. They are scoping out an eight-month mission starting in January 2017, followed by another eight-month workout in January 2018.

Binsted notes that each analog has strengths and weaknesses. "We're in a very physically Mars-like environment. On the flip side, if your study is about a feeling of mortal danger, we don't have that. Our crews know that we can get them to a hospital pretty fast if need be. If you want mortal danger go to Antarctica."

How does it feel, being a principal investigator on HI-SEAS?

"It's kind of stressful, I have to say. My phone stays on 24/7," Binsted replies. "I wake up sometimes worried about something wrong with the habitation module or the volcano decided to erupt. Knock on wood . . . nothing like that has happened. It's personally stressful and it's stressful for the crew too, as it is meant to be. But as we like to say, it's all data."

Only so much like Mars

Today, wannabe Martians are working here on Earth in analog sites that approximate Mars. But no place on our planet presents the weather, geology, atmospheric conditions, and other challenges humans must surmount on the Red Planet. Mars is a maverick world. It has the land area of all the continents of the Earth put together—a place of huge canyons, sand dunes, and towering mountains. Yes, that all adds up to a visual feast—but also treacherous terrain. Tumbling boulders, collapsing lava tubes, and ice caves, as well as Martian windstorms, add up to dangers that Mars explorers will come across.

Supporting the first humans from Earth in the unearthly scenery that is Mars is a work in progress. Engineers are busily sketching out what home base on Mars might look like. Initial habitats are likely to be meager live-in affairs. Over time, however, use of three-dimensional printing and fabrication technology could shape an early home base, and the Mars neighborhood will soon be expanding. ■

FAR AWAY AND COLD ALL YEAR

From the Halley Research Station—established 50 years ago by the British Antarctic Survey on the Brunt Ice Shelf—come insights into the long-term impact of isolation on human behavior, health, and well-being. Halley VI, shown here, opened in 2013.

EARTH ANALOGIES

Medical doctor Beth Healey (right) observes the effects of extreme living conditions at the French-Italian Concordia station in Antarctica, where crews operate in subzero temperatures outdoors and close quarters indoors. Some maintain a sense of humor, as shown by the igloo (below) a winter crew built in 2013 to greet the incoming summer researchers.

CLOSER TO HEAVEN

Nothing impedes the view of the Aurora Australis from Concordia, though witnesses must tolerate extremes, such as an average temperature of minus 60°F. Research ongoing at the French-Italian station focuses on Earth's glaciers and atmosphere, as well as the human response to living conditions as stark as those on Mars.

SO NEAR AND YET SO FAR

In August 2015 six scientists moved into this solar-powered dome on Hawaii's Mauna Loa for 365 days of isolation as part of NASA's HI-SEAS project simulating life on Mars. "Even though it does not feel like living on Mars, we certainly feel very, very far away from the rest of humanity," says Christiane Heinicke, chief scientific officer of the project. Occasionally members emerge, as shown here, on an EVA—extra-vehicular activity.

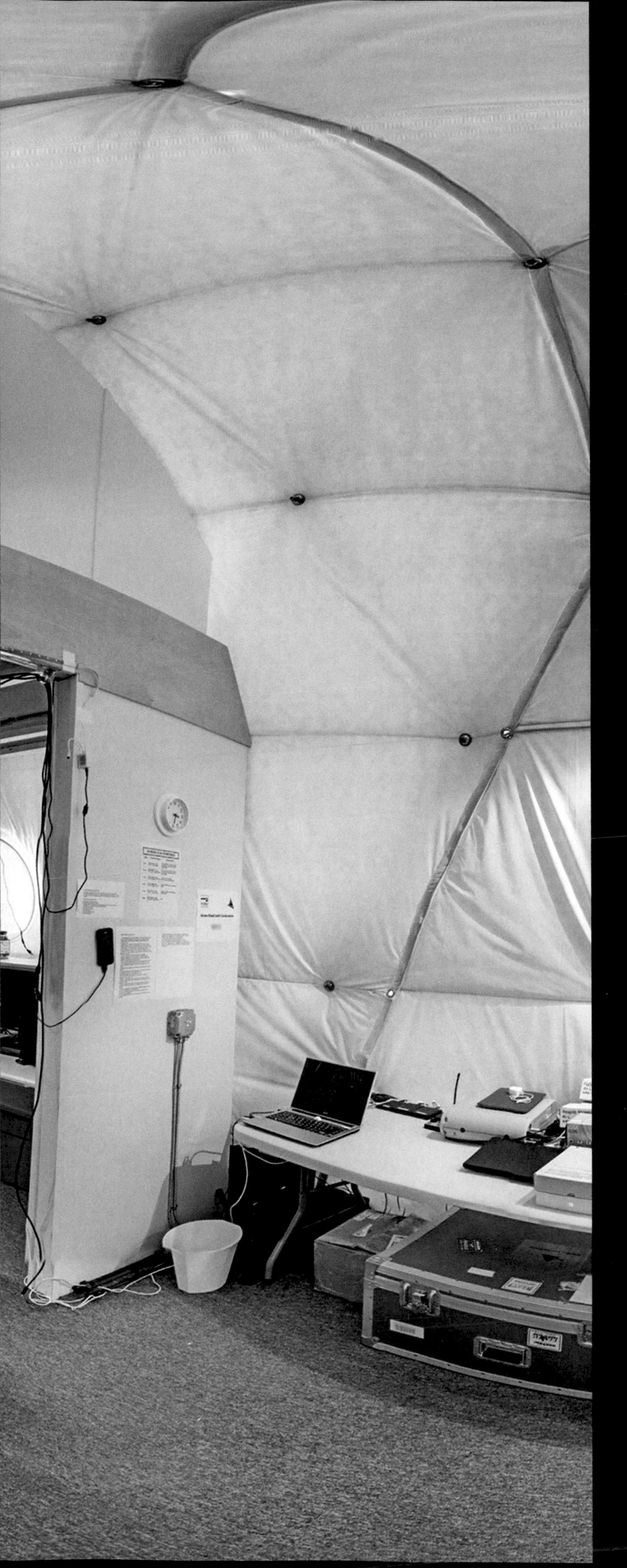

COZY QUARTERS

Residents of the HI-SEAS dome call their home "sMars"—simulated Mars. A geodesic dome 36 feet in diameter, it includes a 993-square-foot common area with a kitchen and a 424-square-foot loft, divided into six private rooms. EVAs include scouting trips to investigate lava tubes (below) in the barren volcanic mountain terrain.

EXIT
OVHD
PORT

HEROES | NICK KANAS

Professor Emeritus, Department of Psychiatry, University of California, San Francisco

Before expeditions can make their way to Mars, a variety of psychological, psychiatric, and psychosocial questions merit consideration. According to Professor Nick Kanas, the accounts of astronauts and cosmonauts who spotlight the stress and strains of space travel are most constructive in gaining this knowledge. Kanas, emeritus professor of psychiatry at the University of California, San Francisco, and a leading specialist on psychological issues related to a trip to Mars, has been the principal investigator of two large NASA-funded studies involving the Russian Mir outpost and the International Space Station. These studies have gained insights helpful for training astronauts to deal with psychological stressors in space.

"It was very important to have multiple missions and a lot of bodies," Kanas says, to accumulate a wealth of data on astronauts, cosmonauts, and mission control subjects. In the last 10 to 15 years, he notes, there has been more variability of individuals taking space trips—breaking the early mold of male test pilots exclusively.

A deep space mission of humans to Mars is rampant with psychological concerns. "You are really going to be isolated. There's no way home if you have a problem on Mars . . . no way of sending somebody back quickly to Earth with a physical or mental problem. You are going to have to deal with it there," Kanas points out.

Kanas and colleague Dietrich Manzey, a professor of work, engineering, and organizational psychology at the Berlin Institute of Technology in Germany, have suggested that Mars crews may experience what they coin as the "Earth out of view" phenomenon. "It's hypothetical. Nobody knows if it's real," says Kanas. "The one thing that astronauts have viewed as a positive is seeing the beauty of our planet in the heavens, realizing how important it is," whether they are seeing it from Earth's orbit or looking back homeward from the moon.

What if you take that view away? At times, in the worst case, a person standing on Mars would be unable even to spot Earth due to the position of the planets relative to the sun. Toss in not having near real-time, back-and-forth chitchat between friends, family, and Mission Control compatriots. "The Earth is no longer a beautiful thing—more an insignificant dot," Kanas continues, "plus you can't talk to anybody in real time."

The Earth-out-of-view phenomenon may give people the sense that everything that's important to them is inconsequential, Kanas adds. Or vice versa, it could highlight a sense of isolation, being distant and away from everything. "It's a different sort of state. Whether that will produce depression, or psychosis, or extreme homesickness . . . I don't know. We have a lot of questions that Mars is going to raise, and we don't have the answers."

NASA astronaut Kjell Lindgren trims the hair of Russian cosmonaut Oleg Kononenko, seated, inside the International Space Station's Harmony module. The clippers he's using include a vacuum attachment, necessary to keep the loose hairs from floating free.

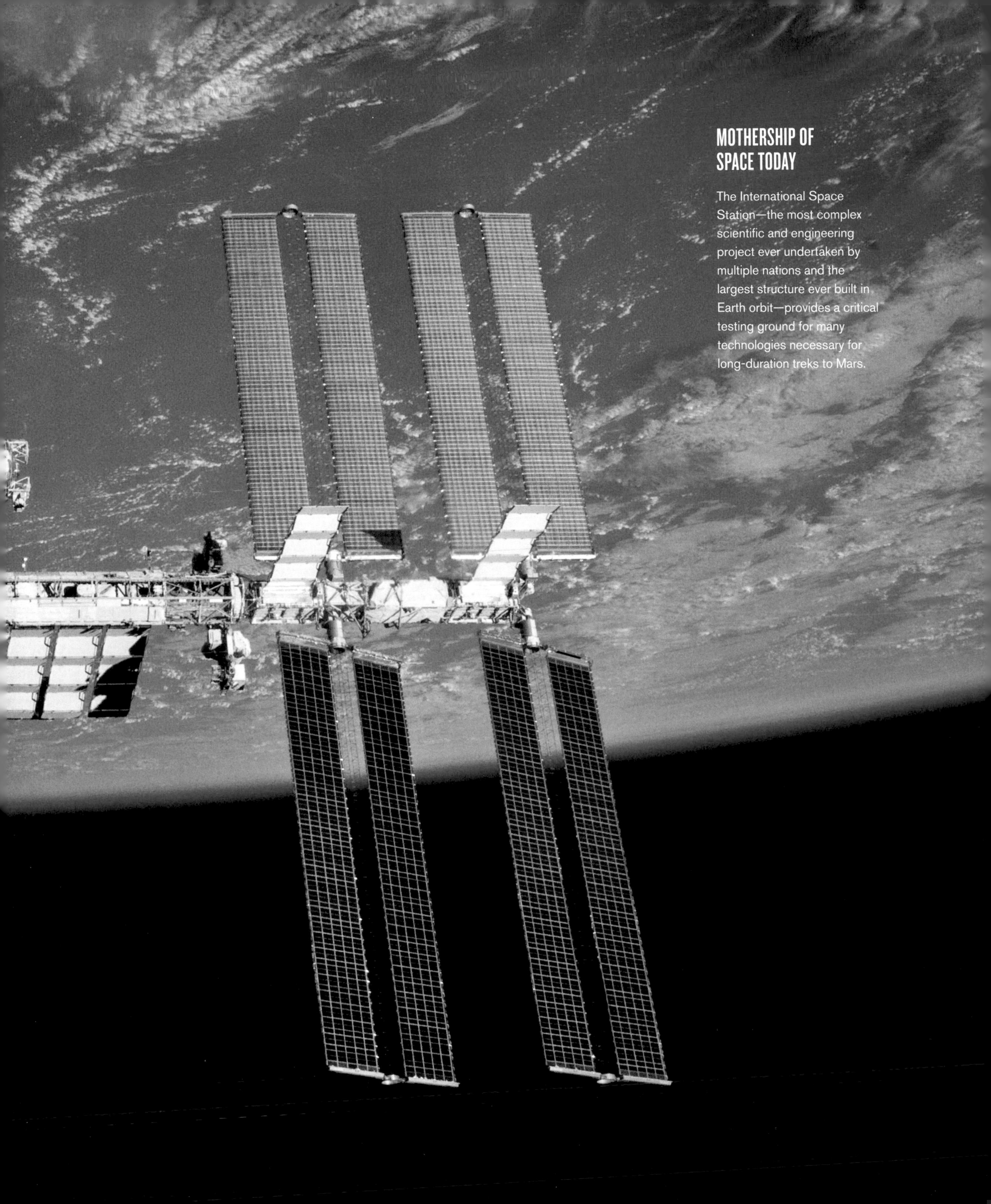

MOTHERSHIP OF SPACE TODAY

The International Space Station—the most complex scientific and engineering project ever undertaken by multiple nations and the largest structure ever built in Earth orbit—provides a critical testing ground for many technologies necessary for long-duration treks to Mars.

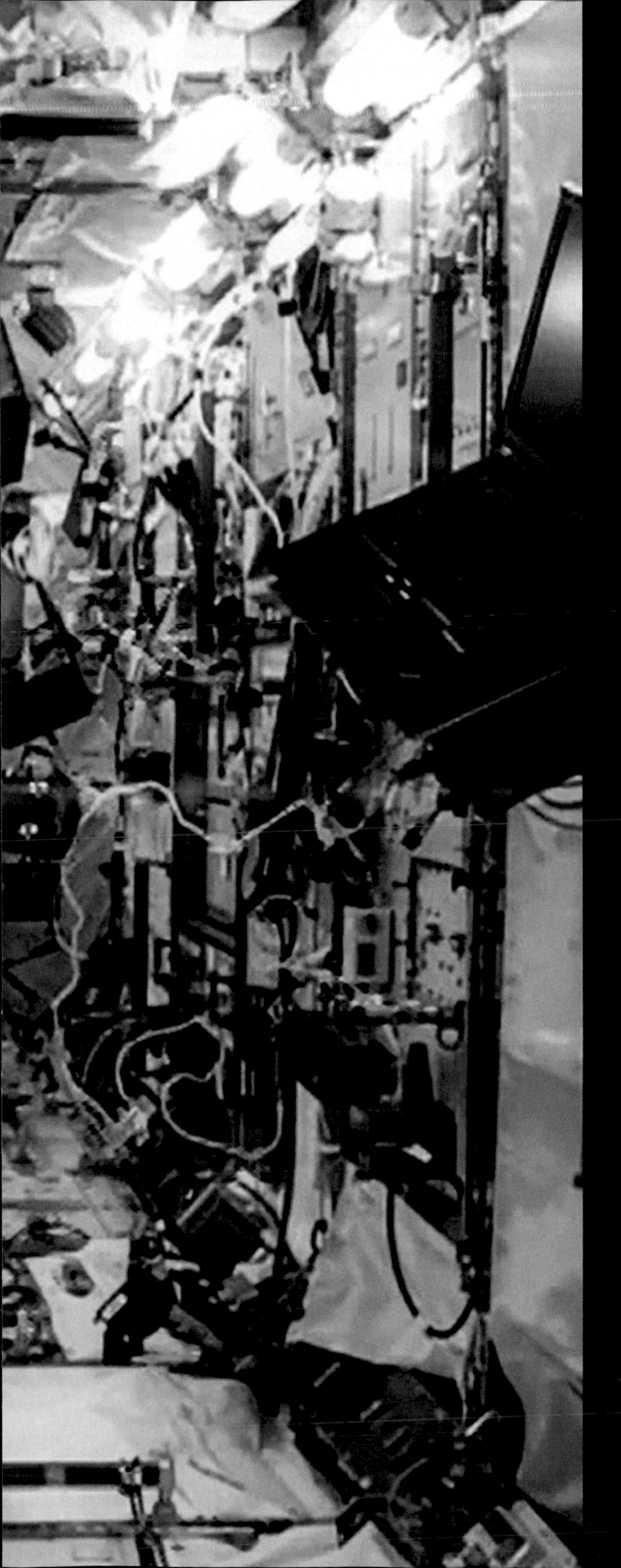

JAZZ AND JAVA IN ORBIT

Canadian astronaut Chris Hadfield (opposite), aboard the International Space Station in 2012–13, charmed the world with his YouTube rendition of David Bowie's "Space Oddity." Two years later ISS residents, including European Space Agency astronaut Samantha Cristoforetti (below), sipped brew made from a newly designed "ISSpresso" machine, which doubles as a coffeemaker and an experiment in fluid movement at zero gravity.

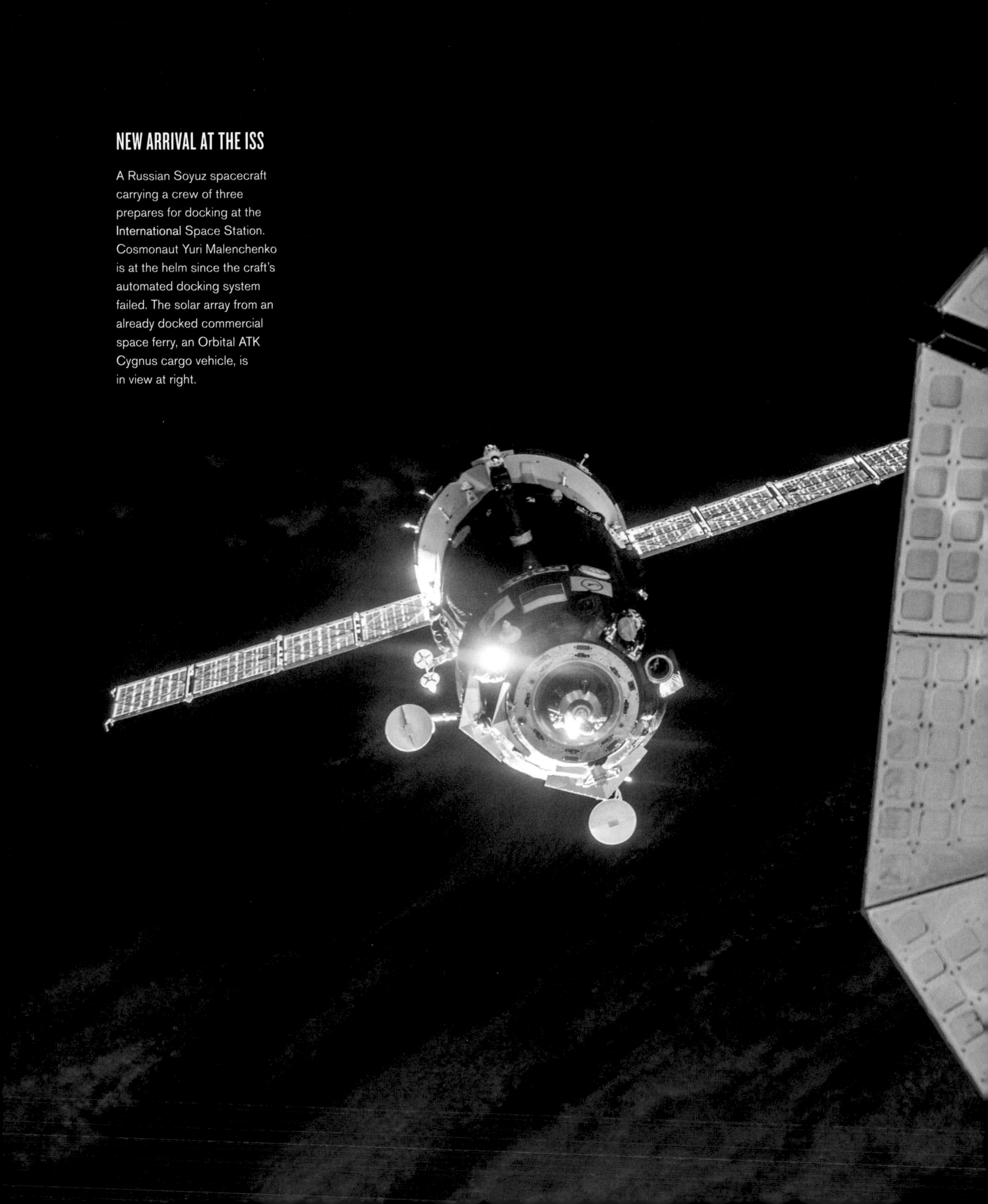

NEW ARRIVAL AT THE ISS

A Russian Soyuz spacecraft carrying a crew of three prepares for docking at the International Space Station. Cosmonaut Yuri Malenchenko is at the helm since the craft's automated docking system failed. The solar array from an already docked commercial space ferry, an Orbital ATK Cygnus cargo vehicle, is in view at right.

WELCOME HOME PARTY

Technicians and press swarm the Soyuz TMA-18M spacecraft just after its landing in a remote area near the town of Zhezkazgan, Kazakhstan, on March 1, 2016. Inside, American astronaut Scott Kelly and Russian cosmonaut Mikhail Kornienko had just completed nearly a year in space.

S. KELLY
КЕЛЛИ
NASA

HEROES | MARK KELLY AND SCOTT KELLY

NASA Astronauts and Engineers

U.S. astronaut Scott Kelly floated back to Earth under parachute in a Russian Soyuz vehicle on March 1, 2016. He had spent 340 days, nearly a year, nestled within the International Space Station, taking part in experiments that will help pave the way to propelling humans to Mars.

During that time, Scott Kelly's Earth-bound twin brother, Mark Kelly, a retired NASA astronaut and engineer, participated in a novel study about how space travel shocks the body. For NASA, the twin study is a new class of work, literally at the genetic level. The Kellys are the only twins to have traveled in space. With an eye to Mars, data were collected on both brothers, which should be useful in identifying potential issues for the lengthy trudge to the Red Planet. Long-duration medical and psychological expertise is fundamental to plotting out a round-trip trek that is likely to last 500 days or longer.

Before, during, and after Scott's flight, the Kellys underwent physical and cognitive tests. While Scott was in orbit and since his return, Mark periodically underwent a battery of blood draws, ultrasounds, and other tests. And after landing, Scott took full measure of his physical state, noting that his stay in space caused his spinal disks to expand, making him an inch-and-a-half taller. "Gravity pushes you back down to size," he says.

Upon his return to terra firma, Scott Kelly said, "I think the only big surprise was how long a year is," adding he benefited by looking at his home planet through station windows. "The Earth is a beautiful planet . . . It's very important to our survival, and the space station is a great vantage point to observe it."

Scott Kelly advises that space travelers need to focus on the tasks at hand. "Take it one day at a time. It's very important. I tried to have milestones that were close, like when is the next crew arriving? When is the next visiting vehicle arriving . . . the next big science activity?" In his opinion, getting to Mars is doable "if it takes two years or two and a half years." The big motivator is to be first to get there. Still, he says, challenges exist, singling out the radiation exposure of getting to the far-off planet.

Participating in the twin study has been a positive experience for the brothers. "As an astronaut having flown four missions while I was at NASA, I would have to say, as far as human research goes on me, this is probably by far the most, collectively, that I've done," Mark notes.

Medical findings aside, Scott made an immediate postlanding conclusion about his brother: "You saw a better tan . . . because he plays too much golf."

Scott Kelly gives two thumbs up just minutes after he and cosmonauts Kornienko and Sergey Volkov landed. Scott was later joined by his twin brother, Mark, both participating in a NASA study on the effects of long-duration weightlessness on the human body.

ENGINEERING A MARS LIFESTYLE

Mars500, a joint project of the Russian Institute for Biomedical Problems and the European Space Agency, has staged several simulated sojourns on Mars in a specially designed facility (right) in Moscow. Engineer Diego Urbina from Italy (below) and five crewmates spent 520 days inside the habitation from June 2010 to November 2011.

DESERT EXERCISES

An isolated, barren landscape makes southern Utah the right spot for simulating Mars habitation. Since 2001 the nonprofit Mars Society has run the Mars Desert Research Station there, designed to create an experience analogous to life on Mars.

DIGGING INTO MARTIAN LIFE

Researchers (left and below) at the Mars Society's Desert Research Station in Utah dress as they will have to on Mars and set out to collect soil samples, using tools like those one would use on Mars. Every activity at the research station provides test results that will apply to life on the Red Planet.

HEROES | JIM PASS

Chief Executive Officer, Astrosociology Research Institute

It's easy to get caught up in the technological marvels of space travel. And no wonder. The mastery of space is hard-core, high-tech engineering to the max. But Jim Pass, of California's Astrosociology Research Institute, believes it's also important to embrace the scientific study of "astrosocial" phenomena—the softer side of social, cultural, and behavioral patterns related to outer space travel.

Since the space age began more than 50 years ago, the emphasis has been on the STEM subjects—science, technology, engineering, and math, says Pass. "The 'S' in the acronym stands for science, but it does not include the social and behavioral sciences, let alone the humanities or the arts." Today, people talk of STEAM power, which adds in the arts. And while that's a good step forward, he says, it still doesn't encompass the social sciences. "We cannot settle Mars successfully unless there is significant convergence between the two branches of science, and I see astrosociology as a means to make this happen." He adds that social scientists will be needed on Mars just as we need them on Earth.

Migration to other space environments such as Mars and the moon seems to be the course humanity will take, Pass says. And the paybacks are many: mining asteroids for resources needed on Earth, easing overpopulation and resource depletion from overuse, avoiding the extinction of humanity due to a global catastrophe, and fulfilling humanity's yearning to explore new frontiers.

"These and other benefits are important, but we need to go about migrating to Mars in a responsible way," Pass cautions. In particular, he underscores the implication of an enormous communications lag between Earth and future Mars settlers. "This delay . . . means that much decision-making needs to occur among the settlers themselves. Moreover, autonomy also tends to result in increased ethnocentrism. At some point, a settlement may break off from its sponsors on Earth . . . so interplanetary relations is another future possibility that we should investigate now," Pass says.

Settling Mars on a sustainable and large-scale basis will likely take decades, Pass observes, so considering what's next is an exercise that looks at the far future, perhaps 100 years from now. "I can see humanity mining asteroids as well as astrobiologists and planetary scientists investigating various bodies in space such as Europa, and living in space stations at different locations."

That's why "we need to gain this [astrosociological] knowledge now; the same principles and phenomena that we observe on Earth will reappear on Mars," says Pass. "Because it's the same human species . . . it will transfer elements of its cultures and social structural/institutional solutions to wherever it migrates to in the solar system."

Crew members from the space shuttle *Discovery* strike a pose of satisfaction, having made it to the International Space Station. Earth is visible through the window at their feet.

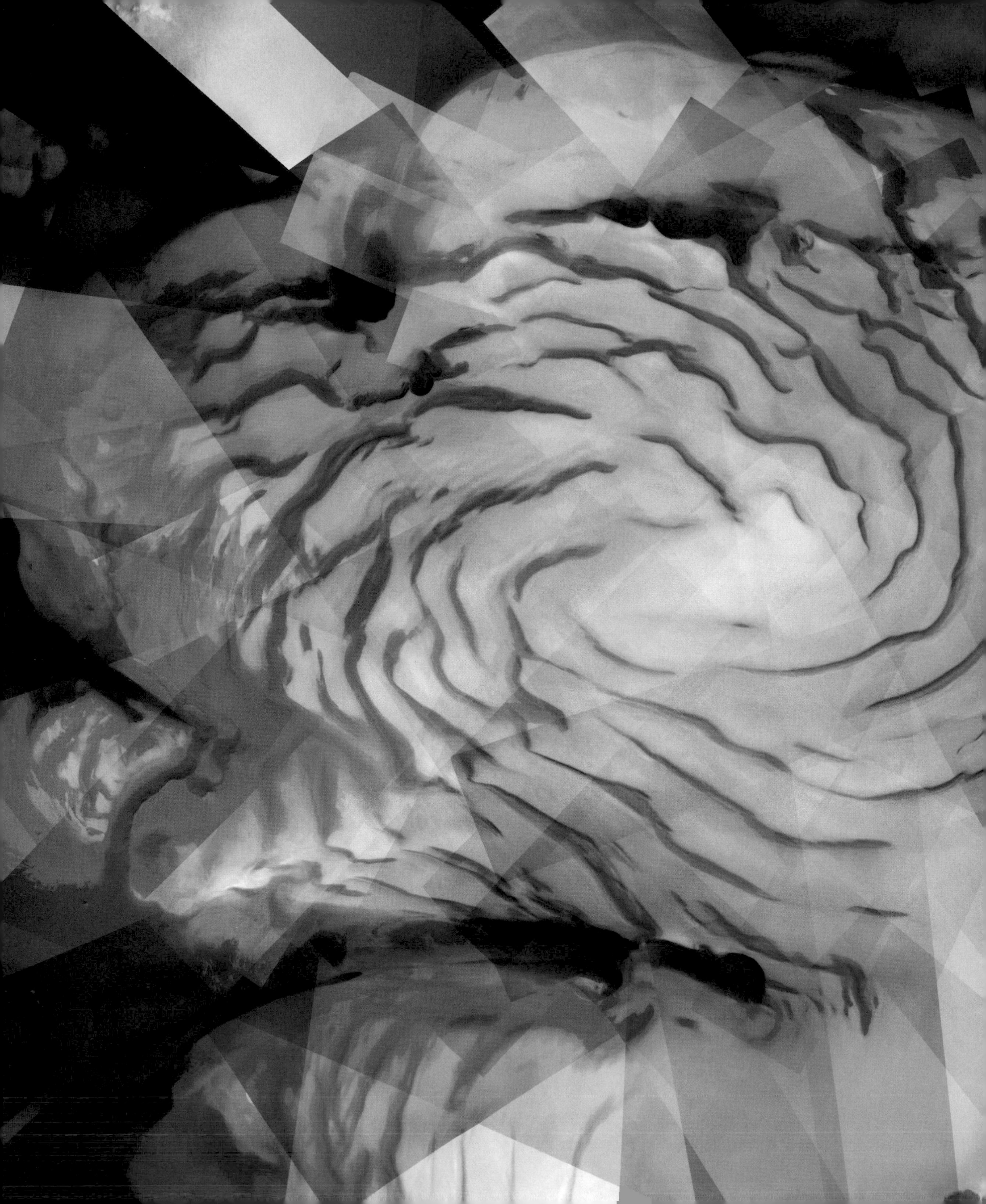

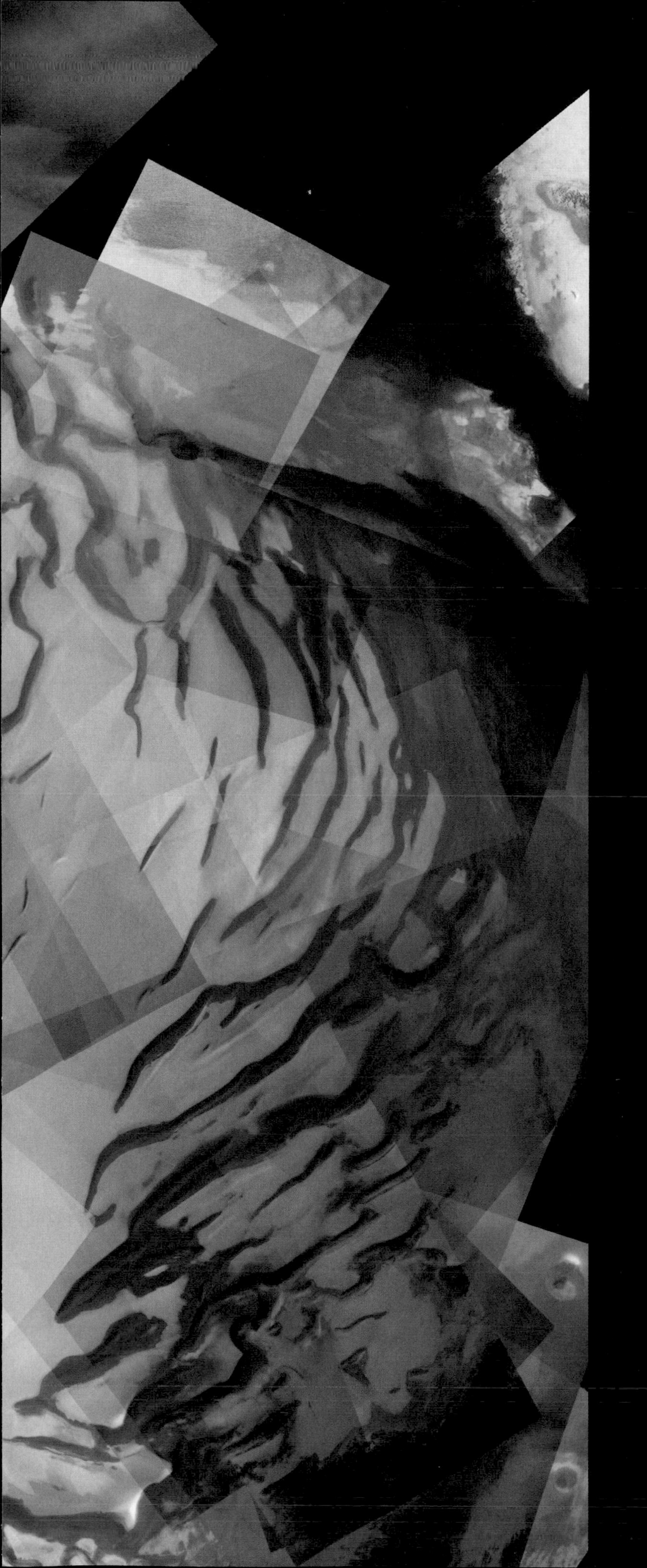

POLAR PERCEPTIONS

Both icy poles of the Red Planet have been portrayed by Mars Express, the European Space Agency's craft orbiting since 2003. A composite of 57 images taken by the mission's high-resolution stereo camera shows the north polar ice cap (left). The same camera captured the south pole (below) in a single sweeping shot taken from 6,000 miles up and calibrated for color and dimensions.

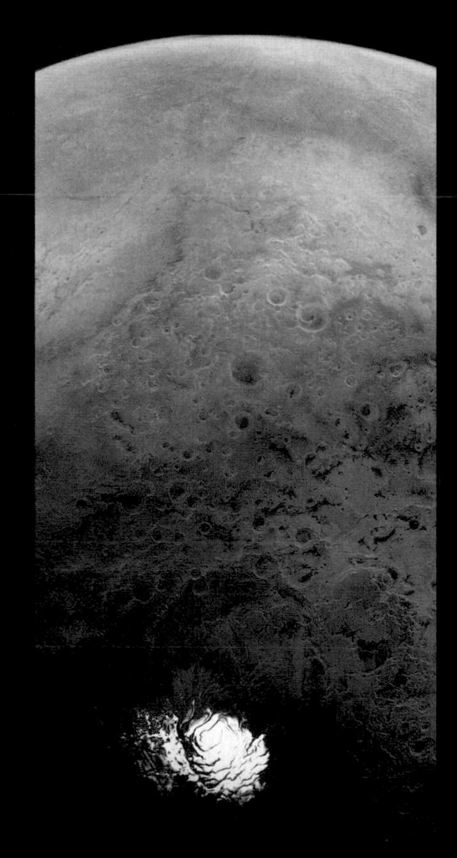

CANDY-COLORED LANDSCAPE

This image of Noctis Labyrinthus, located west of the Valles Marineris, is a mosaic of pictures taken day and night by NASA's Mars Odyssey orbiting spacecraft. Warmer surfaces show up red, cooler surfaces blue. Scientists use this technique to learn more about the geology of a dramatic area such as this.

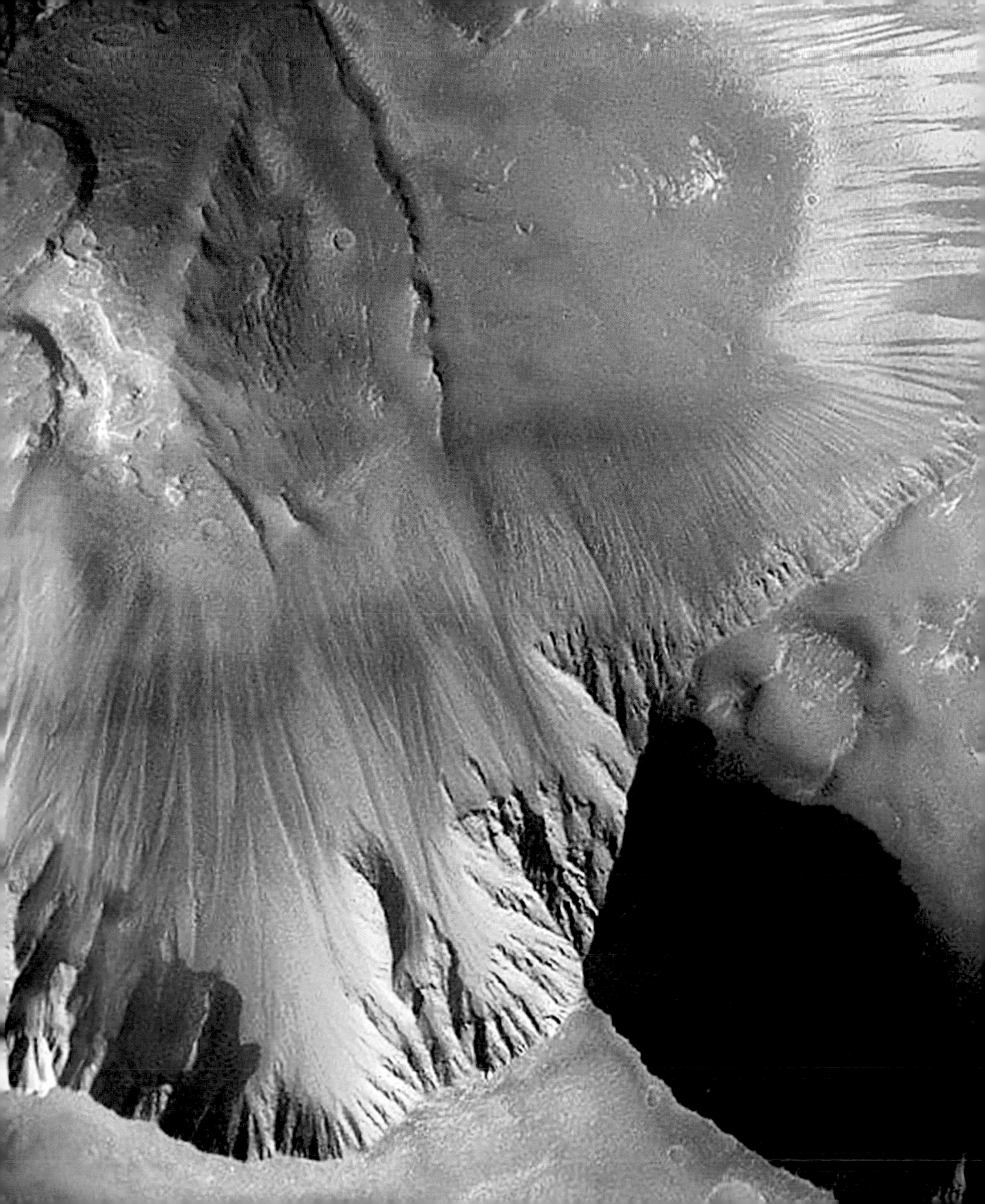

extraordinary demands of huge temperature swings, little or no water, and a death threat of constant radiation.

Victoria Crater, home for more than a Mars year to NASA's Opportunity rover, measures about half a mile wide, with steep cliffs forming its scalloped perimeter and shifting dunes its filigree center.

3

Home Base

MARS IS NOT A REDDISH-COLORED EARTH. It is a complex, ocean-deprived planet. Bone-dry, rocky, and cold, Mars observed from high above or at ground level is an amalgam of features. This world is home to the largest volcano and the deepest canyon known in the solar system.

Surface winds on Mars are typically gentle. Wind speeds have been judged to be about 6 miles an hour, but gusts can be as high as 55 miles an hour. However, the breezes on the planet are timid and wield less force than here on Earth. That's due to the lower density of the Martian atmosphere, which is about a hundredth that of our planet.

And yet Mars has a mean side. Being wispy thin, the Martian atmosphere exposes Mars walkers to lethal levels of space radiation. With no significant ozone layer coupled with a lower total atmospheric pressure than on Earth, Mars presents an environment with a higher surface flux of ultraviolet radiation. Space medical specialists already caution that radiation exposure of hardworking astronauts on the Red Planet puts the threat of cancer right in the line of duty.

Then there are the temperature mood swings of the planet, ranging from 86°F near the equator to a super-chilly minus 284°F near the poles. Top it all off with a coating of perchlorate, a toxic substance that can affect the functioning of the thyroid gland at sufficiently high doses. Not a friendly place.

All of these factors have to be taken into account as space architects and engineers design the structures that humans on Mars will call home. The first order of business is figuring out how to breathe. The air on Mars is only one percent as thick as on Earth, equivalent to our air at 100,000 feet altitude, and it consists of 95 percent carbon dioxide (CO_2). There is almost no oxygen in its pure form, a challenge at the heart of work under way by Michael Hecht of MIT's Haystack Observatory.

As the Curiosity rover made its way to Mount Sharp in 2015, it captured this view of the rocky terrain it was negotiating. Strata here provide evidence suggesting water flowing in eras past.

EPISODE 3

PRESSURE DROP

The first human mission to Mars is in jeopardy. A series of problems stemming from *Daedalus's* troubled descent have left the Mars team behind schedule and struggling to maintain even basic living quarters. As the controlling organizations back on Earth begin to question the value of continuing the mission, the team must race against the clock to locate a suitable site for their settlement—or the next rocket to come to Mars will end their mission and take them home.

"Fortunately, however, there are two oxygen atoms in each carbon dioxide molecule," he says, "and with sufficient power, it is possible to use that CO_2 to make oxygen gas. In fact, trees do it all the time!" A system designed to replicate that process is to be on board NASA's Mars 2020 rover mission: a specialized reverse fuel cell, the Mars oxygen in situ resource utilization experiment, nicknamed MOXIE. MOXIE collects the natural CO_2 found on Mars and splits it into carbon monoxide (CO) and oxygen (O_2) molecules by electrolysis. If proven to work on the Mars 2020 mission, a MOXIE-like system could crank out breathable oxygen for crews on Mars and produce liquid oxygen as a rocket fuel for their return trip to Earth.

MOXIE is a first-of-its-kind experiment, Hecht says, that draws attention to "the art of using what nature offers to substitute for things we would otherwise have to bring along with us." Next question: Can nature on Mars provide water adequate for human life?

NASA's Mars Reconnaissance Orbiter has been circling the Red Planet since 2006, profiling the globe in great detail with its high-resolution imaging science experiment (HiRISE) camera system, the largest ever carried on a deep space mission. The results are gorgeous—and scientifically significant.

"HiRISE has transformed Mars into a familiar world," explains Alfred McEwen, professor of planetary geology at the University of Arizona in Tucson and principal investigator for the superpowerful camera. Those HiRISE images resolve features a person could see while hiking across the surface, he says, and "have shown that current processes on Mars can be strange . . . like gullies carved by particles fluidized by seasonal carbon dioxide frost."

Those fresh gullies take on the look of aqueous features here on Earth. Mars experts in 2014 offered the strongest proof of sporadic flowing liquid water present on the planet, which might help sustain expeditionary crews on the planet. "Finding significant quantities of water on Mars is a game changer," says Rick Davis, assistant director for science and exploration in NASA's Science Mission Directorate in Washington, D.C. Nevertheless, finding the faucets on Mars adequate in volume to support human habitation is a work in progress. What does it take to produce exploitable water from Martian feedstock, be it in the form of ice, hydrated minerals, or from deep underground aquifers? "We need to get a lot smarter," Davis advises, to pin down where best to place a base on Mars for the best access to water.

Where to put down roots

Dozens of places on Mars are under serious scrutiny as a future site for the first human encampment, all within 50 degrees north or south of the Martian equator—the region, as on Earth, that will be the warmest. To be designated an early exploration zone, a site must satisfy several criteria. It will need to provide a setting for three to five landings and support a home base for a four- to six-person expeditionary crew on an expedition lasting about 500 sols, almost an Earth year and a half. It should provide reasonably easy access to scientifically attractive areas and potentially exploitable resources.

NASA's Larry Toups and Stephen Hoffman of Science Applications International Corporation in Houston, Texas, humans-to-Mars planners, see an early Mars station having zones of action. The idea is to keep some activities in relatively close proximity to the station and to distance the station from dangerous events, such as takeoffs and landings, which might throw off dangerous debris from roaring rocket engines. Toups and Hoffman foresee a layout with four discrete zones:

Habitation Zone. This sector would be the heart of the Mars station, with the crew habitat, research facilities, logistical storage site, and a crop-growing facility.

Power Zone. A Mars station may well be energized by nuclear power systems, although solar energy is possible. If nuclear, the hardware should be isolated from the crew and other infrastructure.

Primary Lander Zone. This area is mainly a landing and departure site for the Mars ascent vehicles. Ultimately vehicle propellants could be manufactured here.

Cargo Lander Zones. Located closer to the habitation area, these zones are transport areas for incoming freight.

Robots on a mission

MARS EXPLORERS WILL USE a diverse set of machines to extend their reach beyond the immediate home base. Surface-launched gliders, instrument-toting balloons, robot snakes, and crawlers to inspect subsurface alcoves such as lava tube caves—all of these robotic devices are under design consideration as machines that can do some of the exploring for human explorers.

In one scenario, sensor-laden "tumbleweed craft" are wind-propelled across the terrain using minimal power. These low-cost vehicles would randomly roam the Red Planet on lengthy sprints, perhaps rolling over the residual Martian ice cap. Capable of executing astrobiology assignments, automated tumbleweeds can also aid in reconnoitering miles of Mars to survey for natural resources.

Robotic "hoppers" may jump from one site to another, with each bound studying the scene and then soaring to another area. The long-lived gear would repeat kangaroo-like hops hundreds of times by "sucking up to Mars"—that is, knocking back quantities of the carbon-dioxide-rich Martian atmosphere for use as propellant. Each hopper uses stored heat from an onboard radioisotope power source to ignite the fuel; the vectored blast shoots the machine to a new landing place.

Attention has been given to balloons lofted from an outpost. Designed to steer themselves, an aerial flotilla can perform detailed appraisals of Mars. Capable of long-duration flight, this approach permits pole-to-pole exploration, as well as targeted reconnaissance. Balloons could place a small rover, a miniature geochemical laboratory, or a small navigation beacon at a place of interest.

Still another aerial ambassador for Mars explorers is the entomopter, somewhat like a robotic bug. These insect-like contraptions flap their wings to patrol the planet. Once unleashed, using the planet's low atmospheric density and gravity, they do their science business, be it snapping images of territory below or gathering samples, and then fly back to their point of departure for unloading, refueling, checkout, and reflight.

Creature comforts

SHELTER FOR EARLY LANDING PARTIES on Mars will be frugal but can evolve over time . . . and in a big way. One notional look at the first outpost designs calls it "commuter" architecture. Habitation modules will be placed in relatively flat and safe locations. Surface mobility vehicles will transport expeditionary teams to nearby regions of greater geological diversity.

Counting on taking a leisurely walk on Mars? Prepare for a lot of prep, advises NASA astronaut Stan Love, who has taken his share of space walks from the

Parachute and balloon technologies may allow soft landings for heavy yet sensitive instruments needed on Mars, as depicted in this vision by the architecture firm of Foster + Partners.

International Space Station. Preparation includes an in-suit oxygen prebreathe to flush the nitrogen out of the body so a space walker doesn't get the bends when the suit pressure is reduced for work outside. The suits themselves need a lot of maintenance, too. "For Mars, we don't yet know how much 'overhead' work the suits will need, nor do we know what atmospheric pressure will be used inside the habitat. That in turn determines how long a person has to prebreathe before going outside," says Love, who speaks from his experience in Antarctica on the ANSMET (Antarctic Search for Meteorites) expedition in 2012–13. "If we can get our Mars suits and equipment to be as easy to use and maintain as modern polar gear, we might expect to spend an average of four hours each day outside the habitat exploring and doing science," he estimates.

To create a habitation and research station on Mars, space travelers will initially repurpose the logistics modules that carried necessary cargo with them to the Red Planet. "Once the basic habitat is there, subsequent crews need to bring their logistics but not habitation," explains Stephen Hoffman, a leading Mars aerospace engineer at Science Applications International Corporation. A cargo module could be repurposed as a plant growth chamber or a stand-alone science module, for example. It won't be big enough to grow food for a full-time crew, but it will be a starting point, big enough to find out what works and what doesn't—and probably big enough to provide a bit of fresh food to garnish the crew's packaged food supply.

"Mars is key to humanity's future in space. It is the closest planet that has all the resources needed to support life and technological civilization. Its complexity uniquely demands the skills of human explorers, who will pave the way for human settlers."

—Robert Zubrin, president of the Mars Society

Abundance of resources

THE PRELIMINARY ENCAMPMENT on Mars is anticipated to grow over the course of multiple human and automated cargo missions spanning upward of two decades. As more and more elements arrive at the site, a steady cadence of increasing infrastructure means that future crews can thrive with fewer supplies shipped from Earth. A phased plan like this has been developed at NASA's Langley Research Center, based on the overarching principle that we "exploit the abundance of Mars resources and do not manage scarcity of resources portaged from Earth." The first two crews will erect a small maze of subsurface habitats with connectivity to storage areas containing vast amounts of fuel, life support fluids, and food. Fuel and life support fluids will be harvested from the Martian top surface, plucked from caches of ice, and taken in from Mars's atmosphere. Wastewater will be recycled for growing some food. Over time, the Mars station will become a proving ground for many new technologies that can lead to independence from Earth and, perhaps beyond, supply fuels, oxidizers, life support, spare parts, replacement vehicles, habitats, and other products for further spacefaring beyond low Earth orbit.

Each subsequent crew arriving at Mars will bring something new to the enterprise, especially with manufacturing and other processes leading toward Earth independence. The goal would be to build rovers completely on-planet using plastics made from Mars resources and metals refurbished from discarded entry, descent, and landing hardware.

The power of 3-D printing—also called additive manufacturing—has already made a difference in space, on the International Space Station. Items have been produced on board the station in a fraction of the time currently required to have such objects manufactured and delivered using traditional ground preparation and launch. So if it works in low Earth orbit, might it work on Mars?

Yes, believes Andrew Rush, CEO of Made In Space, a company specializing in 3-D printing in zero gravity. Additive manufacturing will be an essential cornerstone of sustainable life on Mars, Rush believes. "The fundamental difference between people going on camping trips and those settling the wilderness are the tools settlers take with them," Rush says. Settlers have to take the tools of manufacturing with them. "Manufacturing technologies must accompany the earliest Martian settlers," he says, "and must be upgraded and expanded with each subsequent batch of settlers."

Additive manufacturing devices on Mars will be able to use the planet's own resources to create tools, building materials, food, and more. "Techniques in food printing and many other emerging areas will have evolved significantly in the time between now and a Mars colony," predicts Rush, "enabling, for example, remote manufacturing of Earth delicacies." Meanwhile, ongoing work at the Center for

Rapid Automated Fabrication Technologies at the University of Southern California involves "contour crafting"—automating the construction of whole structures and radically reducing the time and cost of construction. Large-scale parts can be built in layers that are as thick as bricks, thereby allowing rapid construction of large structures. Here on Earth, this approach can turn out high-quality, low-income housing, even the rapid construction of emergency shelters and on-demand housing in response to disasters. Center director Behrokh Khoshnevis sees possibilities for its use on the moon and Mars. The same research center is developing selective separation shaping, using additive manufacturing to fabricate metallic, ceramic, and composite materials out of the resources available on the moon and Mars.

Visionary architecture

So what will Mars lodging look like in the future? Mix accessible resources and cutting-edge 3-D printing with a big portion of imagination, and you can begin to visualize on-the-spot, on-the-planet accommodations. They are literally on the drawing board of many architects, engineers, and planners around the world today.

In 2015, NASA and the National Additive Manufacturing Innovation Institute, known as America Makes, held a competition to recruit creative groups to design a 3-D–printed habitat for deep space destinations such as Mars. More than 165 submissions were received. The first-place winner was Mars Ice House, an igloo-looking beehive structure built entirely of ice. Designed by New York–based SEArch (Space Exploration Architecture) and Clouds AO (Clouds Architecture Office), an architecture and space research collective, the Mars Ice House would be crafted by semiautonomous robotic printers that both gather subsurface water ice and deposit it as the inner and outer walls of the structure. Because they are built with 3-D printing using available materials on Mars, Ice House structures can be completed without bringing heavy equipment, supplies, materials, or structures from Earth.

Mars Ice House draws on the great quantity of water and low temperatures in Mars's northern latitudes to create a multilayered pressurized radiation shell of ice, a structure that encloses a habitat constructed from the lander and gardens but allows light into the living quarters. Even before astronauts arrive on Mars, construction can be accomplished semiautonomously with digital manufacturing techniques. "Ice House is born from the imperative to bring light and a connection to the outdoors into the vocabulary of Martian architecture," the team explains, "to create protected space in which the mind and body will not just survive, but thrive."

Second place in the competition was awarded to Team Gamma, a Foster + Partners design team in New York. Their modular habitat construction approach begins with parachuting onto the Martian surface an array of preprogrammed, semiautonomous robots, arriving well before any astronauts land. Three kinds of robots—diggers, transporters, and melters—go to work to excavate a deep crater, collecting loose soil and rocks along the way. They then fill the hole with inflatable modules and pack the loosened rock and soil back around them, using microwaves to fuse the material into solid walls. The resulting robust 3-D-printed dwelling, 1,000 square feet in size, will support up to four astronauts; the fused Martian soil creates a permanent shield that protects the settlement from extreme radiation and severe outside temperatures. The design combines spatial efficiency with human physiology and psychology, says the design team, with overlapping private and communal spaces, an interior finished with soft materials, and enhanced virtual environments to fend against monotony and create a positive living room setting.

"Our future is cast in lava," claims the team whose design won third place in the competition. LavaHive—designed by artisans from the European Space Agency in Germany and LIQUIFER Systems Group in Austria—is a modular additive-manufactured Martian habitat created by using a unique "lava casting" construction technique and recycled spacecraft materials.

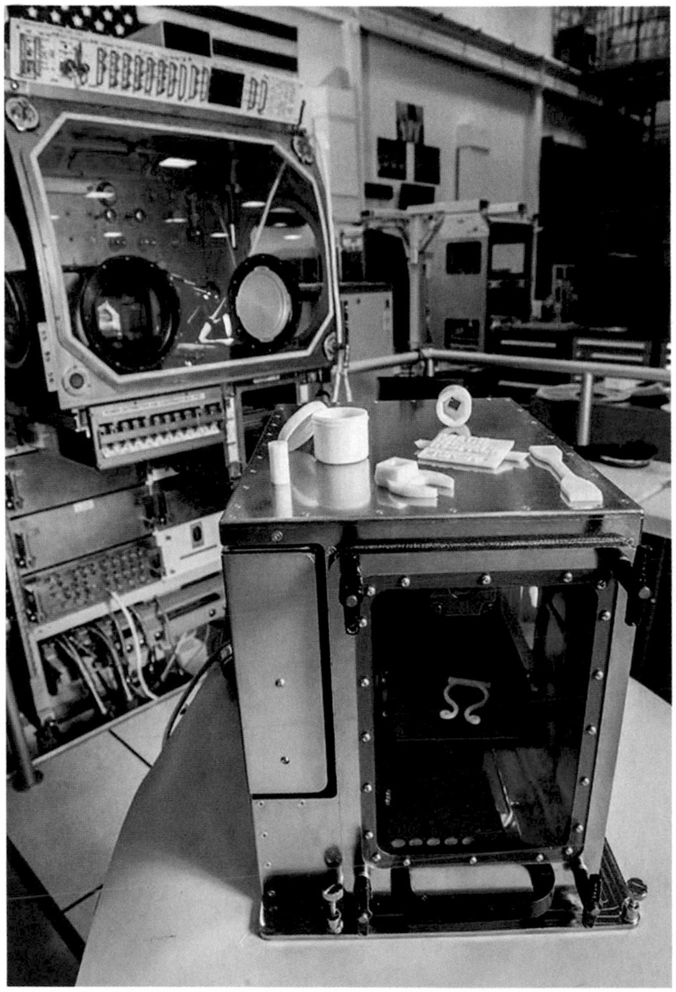

Three-dimensional printing technologies promise on-site manufacturing capabilities for expeditions to Mars. Here the U.S. company Made In Space displays sample printed objects on top of the printer that made them and, behind, the Microgravity Science Glovebox designed to test the process on Earth. The printer now operates in the International Space Station.

The LavaHive habitat starts with one inflatable dome brought from Earth, its roof created from key pieces of the mission's Mars entry vehicle. After that, the crew will mine regolith—loose sand, rock, and sediment from the planet's surface—to be used as a building material. Some will be melted down and poured into forms (the "lava"); some will be sintered, or compressed under heat, into solid structural material. Those components together will be used to build more dome-shaped buildings, all connected. "We envisage using Martian regolith as a building material," says the LavaHive design team leader, Aidan Cowley, "and take this a step further by recycling spacecraft parts that are usually crashed into the planet's surface."

Using this combination of lava-cast and sintered Martian soil, a crew living in the original habitat will be able to build connecting corridors and subhabitats around the main inflatable section. These subhabitats will then be fitted, sealed internally with epoxy, and outfitted as research areas, workshops, or greenhouses, depending on the mission design. An air lock module houses suitports (space lingo for environmentally controlled closets) for the comings and goings of four crew members. The maintenance workshop and docking port can connect to a mobile rover that is pressurized for drives across the Martian landscape. Modular in design and locally sustainable in its construction materials, the LavaHive habitation can be expanded over time.

Native Martian design

DESIGNS FOR HABITATIONS ON MARS must be practical but visionary. For instance, the NEO Native, developed by MOA Architecture of Denver, Colorado, is promised to be "a living shell that responds to its environment and pushes the limitations of not only what we know but who we are," in the words of the design team creating it. Manufactured by 3-D printing using regolith materials as well, this structure will take its shape from the Martian terrain on which it is built, the exterior more reminiscent of a windswept skyscraper lying down on the landscape rather than standing above it. NEO Native designers envision advanced 3-D printing capabilities that can scan the proposed habitation site and then use those specifications to manufacture a structure suited to the environment. They propose to locate NEO Native in the Valles Marineris region of Mars—a site that provides a favorable climate, better potential communication with Earth, and access to billions of years of exposed geology. They compare the site to the sacred Four Corners region of the southwestern United States, home to the Pueblo cultures, where "dwellings provided shelter and protection while serving to establish a cultural identity based on spiritual representation and observation of the earth and heavens." In the same way, explain the NEO Native architects, "As we change the dust and stone of the Martian landscape to the iron and bone of humanity, we must . . . be reminded that, as we observe something older than ourselves, we are also looking into our own impending future."

DISASTERS

Basic necessities | Habitat and clothing may not adequately protect us from solar and cosmic radiation. Technology to supply oxygen could fail. Water on Mars may be inadequate. After we consume the food we brought with us, it's a challenge to grow our own.

What could go wrong?

Some designers are casting their imaginations even further into the future, asking how to create a sustainable living environment for many people, even multiple generations, on the Red Planet. That is the mission of the Mars City Design competition, the brainchild of Vera Mulyani, an architect and filmmaker in Los Angeles, who has high hopes not just for the exploration of Mars but for Mars as a second home for humans. "It is essential that we call on a new generation of thinkers and innovators to make this a reality," she says, "and by using Mars, we may be able to heal Earth

too." Outlining the challenges of building a city on Mars—the brutal atmosphere and climate, the cosmic rays and ultraviolet radiation, low gravity, and the need for self-sustaining materials without excess reliance on Earth—Mulyani and associates have called for innovative design in several categories, from infrastructure and agriculture to human health and services.

Pressurized rovers—essentially roving science labs—will carry Mars explorers and equipment far from a base habitation, thus greatly expanding the exploration zone beyond an initial landing site.

Joining Mulyani in reviewing the designs is a distinguished list of experts, representing how seriously people around the world are taking the question of what life will be like when we go to Mars. Entrepreneur Anousheh Ansari, who funded her own eight-day expedition aboard the International Space Station in 2006, sees this time as "a very historical moment." Gregory Johnson, president and executive director of the Center for the Advancement of Science in Space (which manages the U.S. National Laboratory aboard the International Space Station) foresees "a great challenge in front of us to someday colonize Mars" that will require "all of the innovations and ideas from past space programs and the new ideas and innovations from the next generation." James Erickson, project manager of the Mars Science Laboratory at NASA's Jet Propulsion Laboratory, will review submissions for the Mars City competition as well. "There are constraints on Mars that aren't here on Earth," he says, "but there are constraints on Earth that you won't have on Mars." The time is ripe for blue-sky thinking, says Erickson. "We know that this is the ground floor. We have the opportunity to start fresh." ■

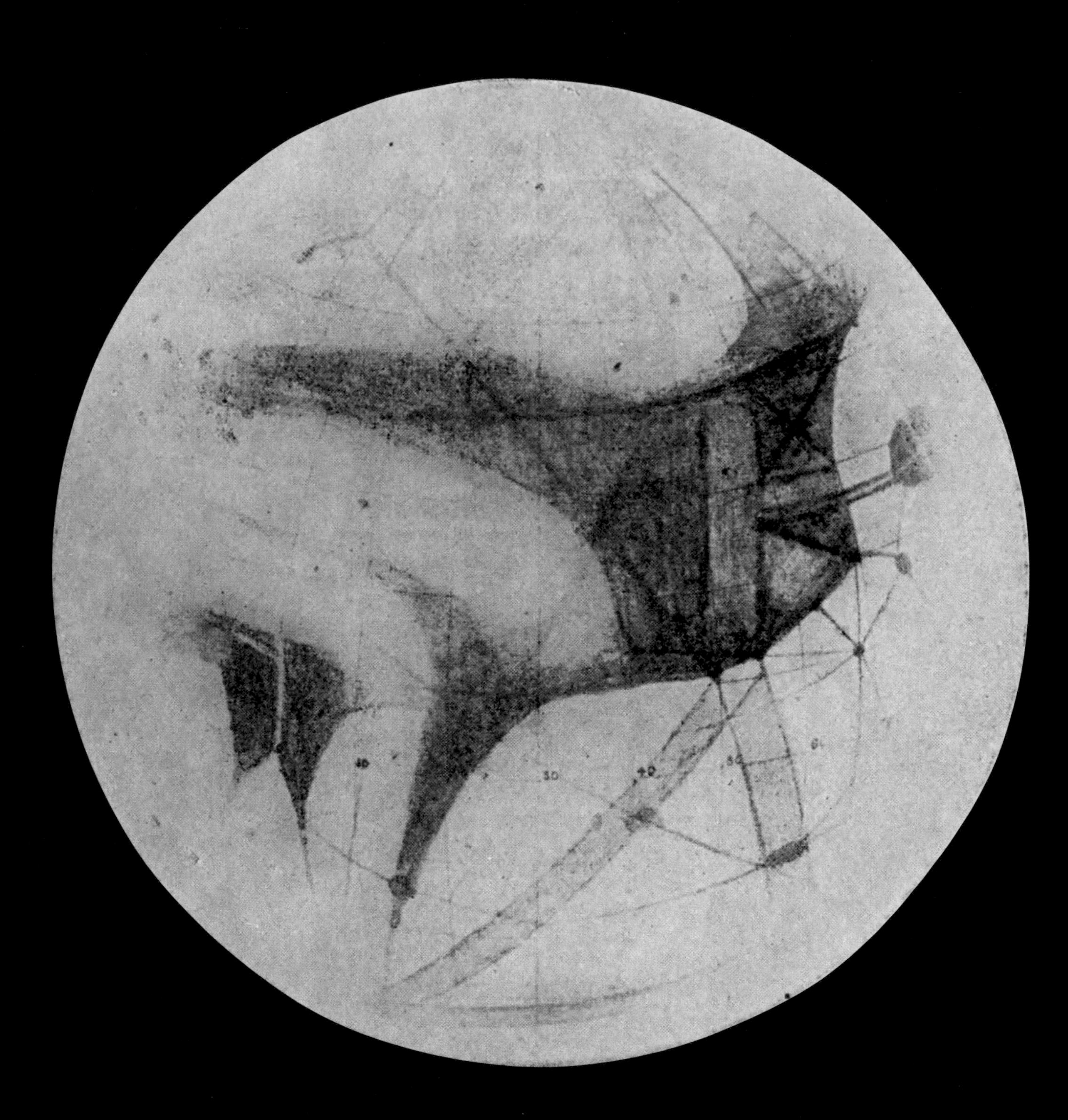

LOWELL'S CANALS

American astronomer Percival Lowell, observing features such as the massive canyon Valles Marineris (opposite), proposed an engineered irrigation system, evidence of "the world-wide sagacity of its builders," in his 1906 book *Mars and Its Canals.* Although his hypothesis now seems unfounded, his extensive records of Mars's changing surface features mesh with recent observations of seasonal change.

TAKING A TUMBLE

Making the most of surface winds, NASA engineers propose "tumbleweeds": instrumented machines capable of moving swiftly across the planet and collecting data.

NASA
NASA
NASA
NASA

EXPLORATION CENTRAL

Every long-term habitation plan—such as this artist's vision of a fully operative Mars research base—involves the slow accumulation of infrastructure, mission by mission. Over time, multiple modules build on an early and simple outpost, cultivating lifestyle and exploration possibilities and extending the stay time for crews.

OCEANEERING

WELL SUITED FOR MARS

Space suit designers face many demands as they engineer next-generation extraterrestrial garb. The PXS (prototype exploration spacesuit, opposite) is more flexible than earlier designs, and parts of it can be produced by 3-D printing. The Z2 (at right), designed specifically for Mars, streamlines sample collection. Each design includes a portable life support system.

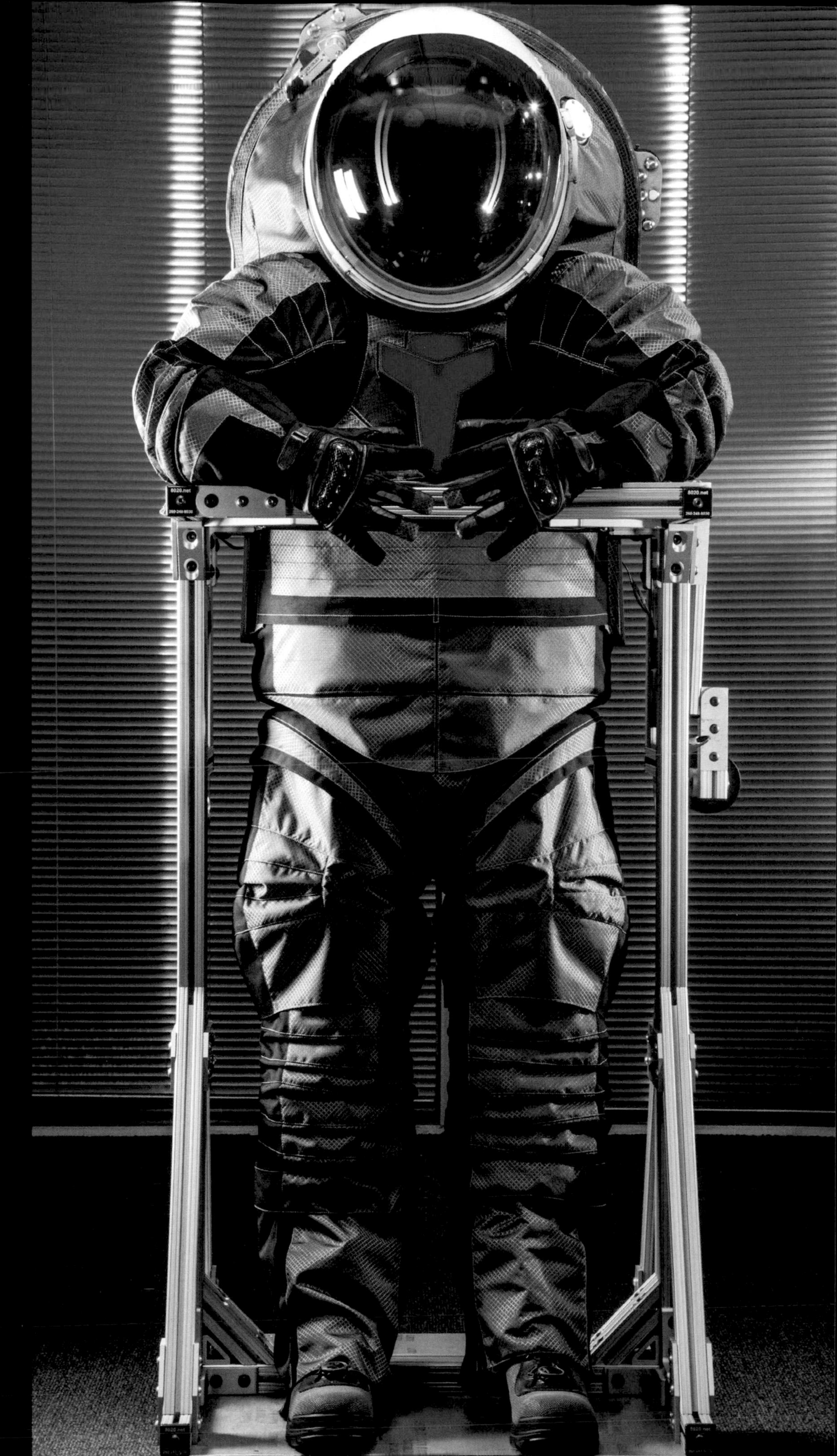

FRONTIER DWELLINGS

Runner-up in a recent NASA competition for 3-D–printed Mars habitats, this Team Gamma design proposes to use local regolith, or surface rock, as the basic material to create a protective shield around a modular inflatable hab.

RAMM
21
RAMM
17
RAMM

6
HUDSON

HEROES | BRET DRAKE

Space Systems Architect, The Aerospace Corporation

Since the 1980s, Bret Drake has been assessing what it will take to get people to Mars and back. A foremost thinker at the NASA Johnson Space Center in Texas, Drake led the Mars Architecture Steering Group that produced the Design Reference Architecture 5.0—a detailed review for placing humans on Mars. He recently departed the space center and is now at the Aerospace Corporation in Houston.

"I've lived through many ups and downs," Drake says. "We know what we need from a systems and technology perspective. It's just a matter of getting on with it, starting to develop those systems, proving them, and moving out." Identified in those detailed NASA reviews were the physical limits of Mars travel. "Those constraints kind of force you into a solution set that we know pretty well," says Drake. "Still, there are things that we do need to figure out like entry, descent, and landing," including the exact technology for transporting crews to the Red Planet.

Over the years, Drake says, there have been changes to blueprinting Mars missions. One of those changes is that missions after the first landing would now return to the same setting on Mars. The intent would be to build up an even larger capability at an explicit site. For subsequent crews, life on Mars would get a bit easier, he adds. "Because the human exploration of Mars is going to be a large endeavor—requiring good effort from lots of nations for many years—a singular mission doesn't make sense."

Today, laying out the route to Mars is complicated. "The moon-huggers want to go back to the moon because they see Mars as too far off," Drake says. "And the Mars huggers don't want to go to the moon because they see it as a distraction that delays humans to Mars." For instance, Europe is pushing the notion of a moon village as a precursor gateway to the Red Planet. Just how that lunar agenda could affect any time line to get to Mars is arguable. He adds, "To find that balance between those competing objectives is challenging."

Drake characterizes NASA's Mars planning today as "progressive expansion." A key step is getting a first-rate launch capability and needed ground infrastructure up and running. Another step, Drake says, is shaking out the Orion spacecraft for long-duration, deep space duties. It adds up to converting near-Earth and cis-lunar space (between Earth and moon, including moon orbits) into a proving ground. These step-by-step demonstrations, he believes, make it feasible to wave goodbye and say good luck to crews headed for Mars.

"It's just knocking down those critical capabilities that we know we need," he says. Making progress on them brings Mars one step closer.

"Mars architecture" means not just buildings but transportation, communication, research goals, and life-support systems tracked over the next few decades of planetary exploration.

NECESSITIES OF LIFE

Early habitations on Mars will expand, allowing crews to extend their exploration zones. Robotics powered by solar cells deliver supplies from a recent arrival to the habitation zone by means of a pressurized rover, powered by solar cells, shown at left.

IMAGINE THE ERUPTION

At 88,500 feet tall and 374 miles wide, Olympus Mons is the largest known volcano in the solar system—so big that, as an artist's interpretation shows (below), it rises as a visible mound from the surface of Mars. The central caldera, almost as wide as the mountain is tall, speaks of massive eruptions and collapses afterward.

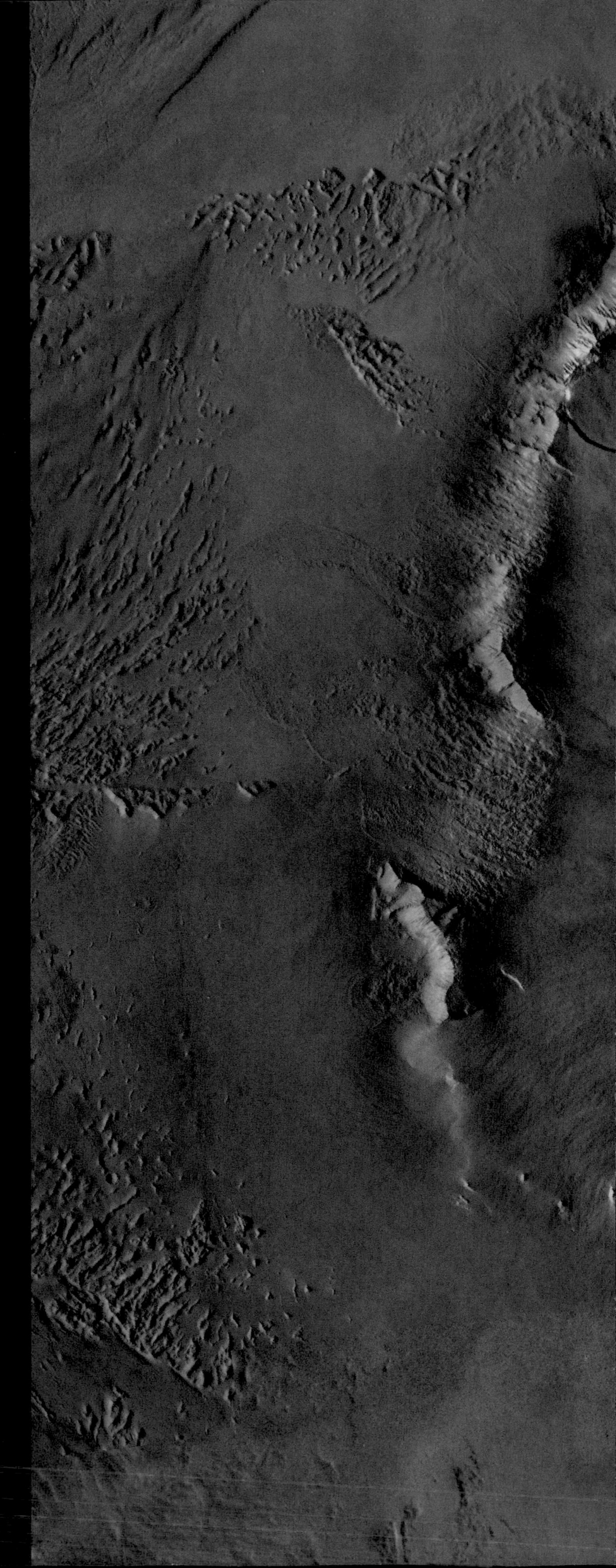

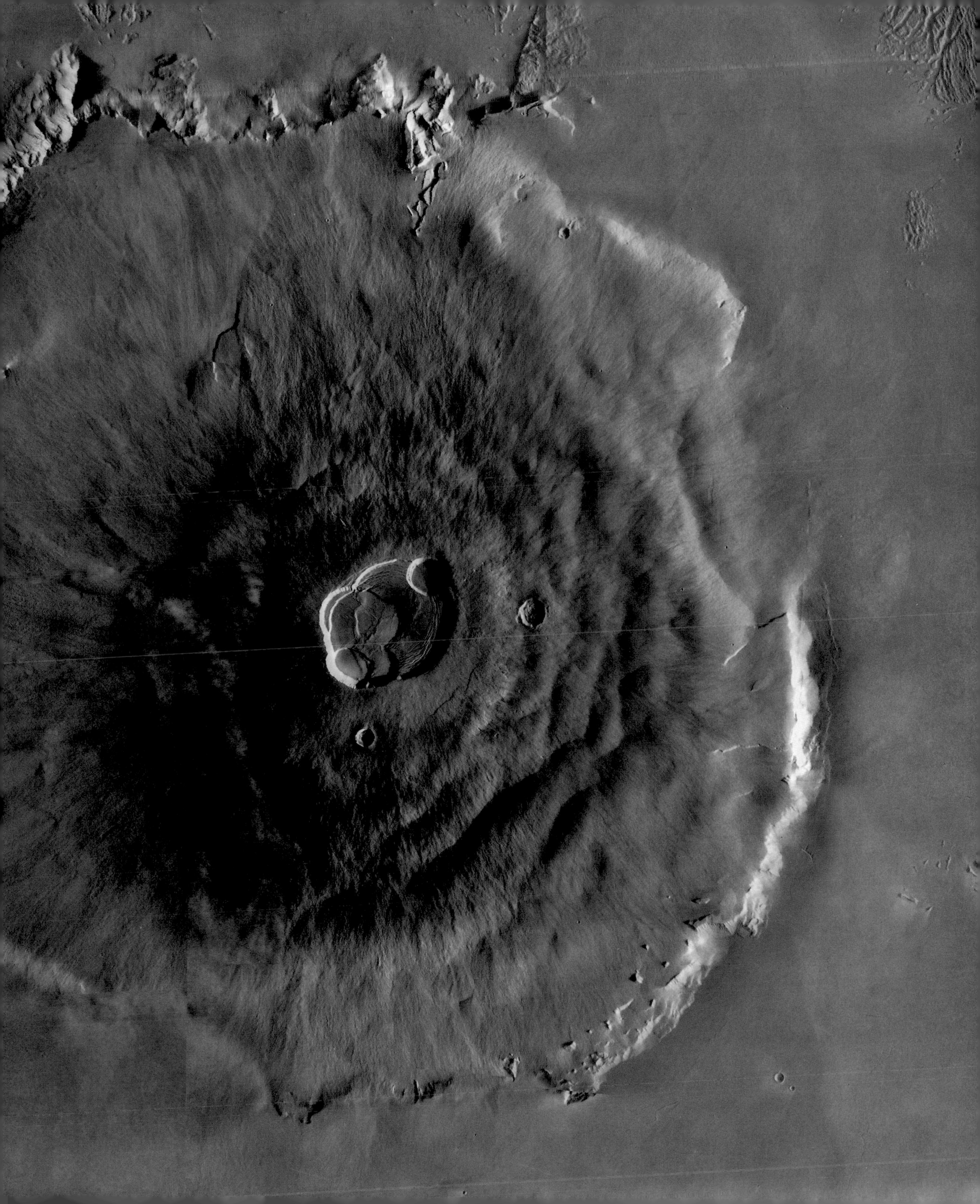

DONUT-SHAPED DECELERATION

An ingenious reentry technology under development by NASA, this 10-foot heat shield compresses into a 15-inch-wide bundle and then inflates with nitrogen into this mushroom shape to slow and soften a payload's landing.

HEROES | PASCAL LEE

Director, Haughton-Mars Project, Mars Institute, NASA's Ames Research Center

Devon Island in Arctic Canada is the largest uninhabited island on Earth and home to Haughton Crater, an enormous impact crater some 23 million years old and roughly 12 miles in diameter. The High Arctic site, set in a polar desert, is remote, stark, and rocky. It's been dubbed "Mars on Earth" because the area's geology and climatology are as close to Mars as can be found on Earth.

"The climate is cold, but not quite as cold as Mars," says Pascal Lee, chairman of the Mars Institute and mission director of the Haughton Mars Project (HMP), an interdisciplinary effort being carried out by the Mars Institute. "The climate is dry, not quite as dry as Mars. The terrain is unvegetated, not completely, but mostly. There is the rocky frozen ground and glaciers."

These features are "in the right direction" of mimicking conditions and landscape on the Red Planet, says Lee, a planetary scientist who has led more than 30 expeditions to the Arctic and Antarctica to study Mars by comparison with Earth. And that's why the island is home for HMP. Future Mars explorers can benefit by traveling to the site, he says, which serves as a welcome mat for revealing safer and more efficient ways to live long and prosper on faraway Mars. Started in 1997, the international field research venture is the longest NASA-funded research project at the surface of the Earth, Lee says.

The project's research station, a cluster of habitats, is a model for how a Mars outpost might be configured and operated. "We've had astronauts visit the place, and we expect more will come as part of their actual training," Lee says. "HMP is a real field exploration setting," he adds, flush with ways to inform the planning and optimization of future human science and exploration activities on Mars.

Furthermore, over the years, HMP expeditionary teams have evaluated all manner of equipment: new robotic rovers, space suits, drills, and aerial drones. Also on the hardware menu are the Mars-1 and the Okarian Humvee rovers—HMP's two simulation pressurized rovers for long traverses into the wilds of an already wild Devon Island. In addition, personal all-terrain vehicles for short treks from base camp are used.

Going into its 20th season, HMP's field campaigns now include international teams of scientists. As future humans-to-Mars plans continue to jell, Lee says, the experiences gleaned from Devon Island will be invaluable. "I see Devon Island as a training site for astronauts bound for Mars," says Lee. "It will become one of the essential stops, if not the final stop, in readying crews for the Red Planet."

Capitalizing on the Mars-like conditions of Earth's only impact crater in a polar desert—Haughton Crater on Devon Island, Nunavut, Canada—researchers on an international interdisciplinary field research project located there have investigated space suits, robotics, and geological sampling, shown here.

HOMESTEADING, MARTIAN STYLE

Visions of permanent human settlements on Mars all include methods by which inhabitants can grow plants and produce food. Greenhouses will need oxygen and water provided, as well as adequate sunlight and temperature control.

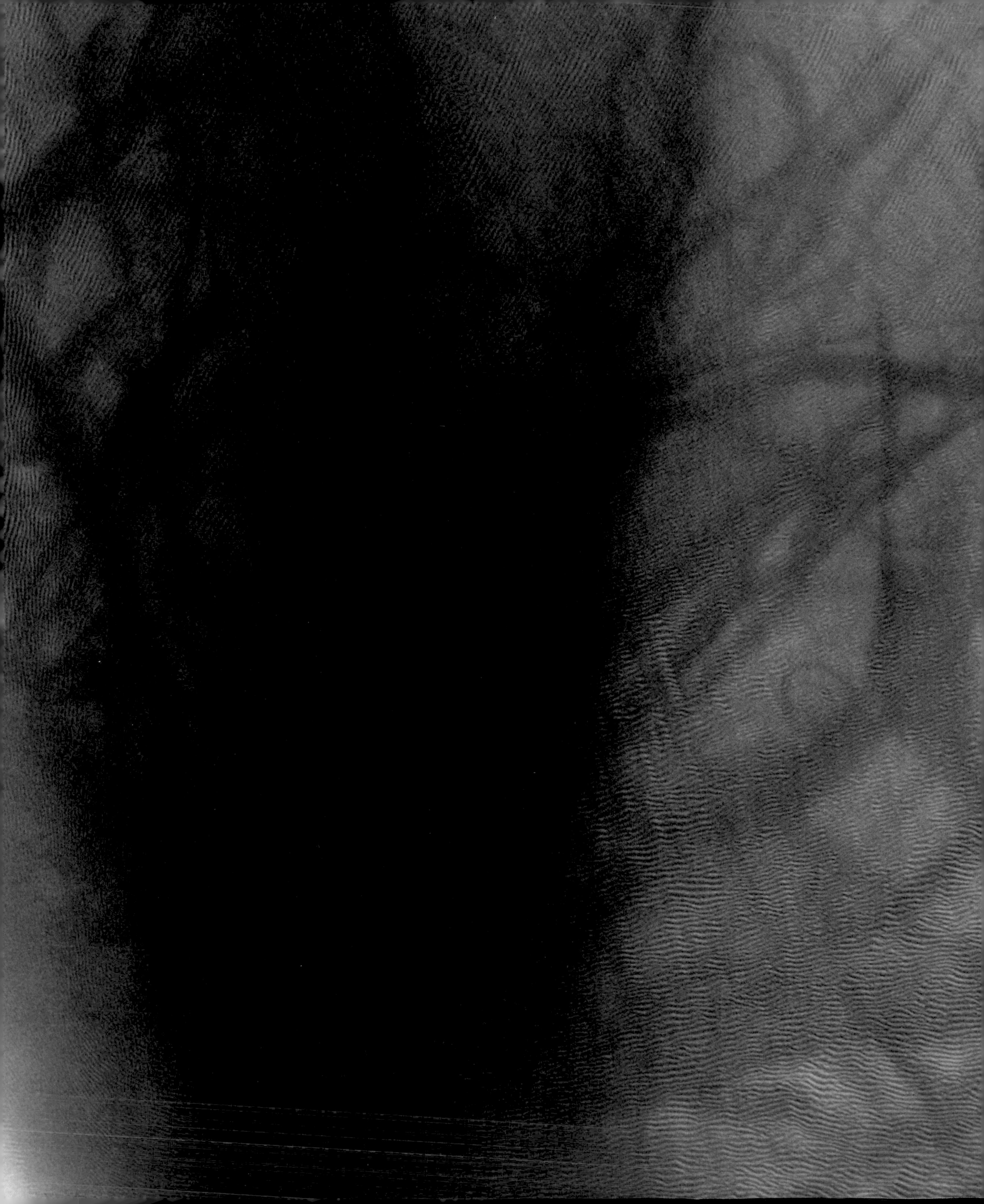

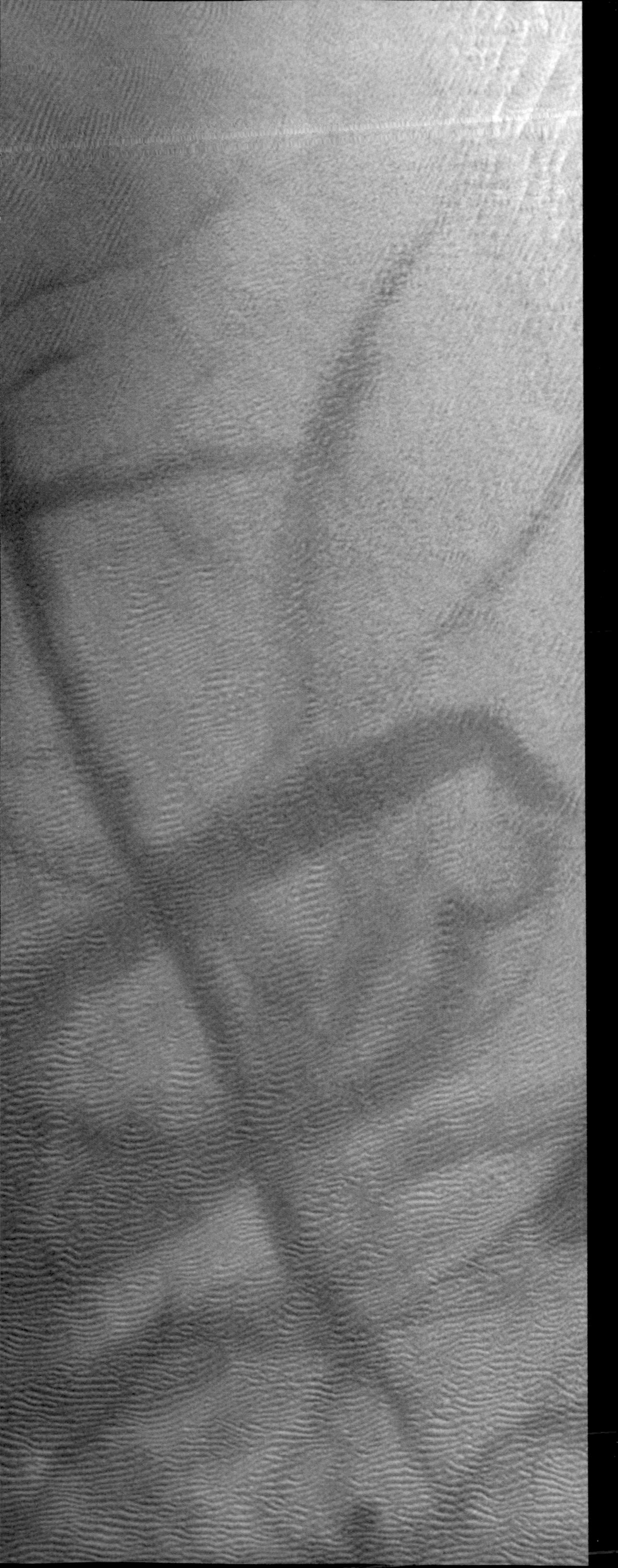

DEVILISH DUST

Dark lines crisscross some areas of Mars, the tracks of dust devils (left). These whirlwinds pick up the light-colored dust on the planet's surface, exposing darker rock material below. HiRISE, an orbiting camera, caught a picture of one especially tall dust devil in 2012 (below). Calculations based on the shadow suggest it stretched half a mile up, nudged serpentine by winds at different altitudes.

At the heart of every Martian enterprise is the question: Is there, or has there ever been, life on Mars?

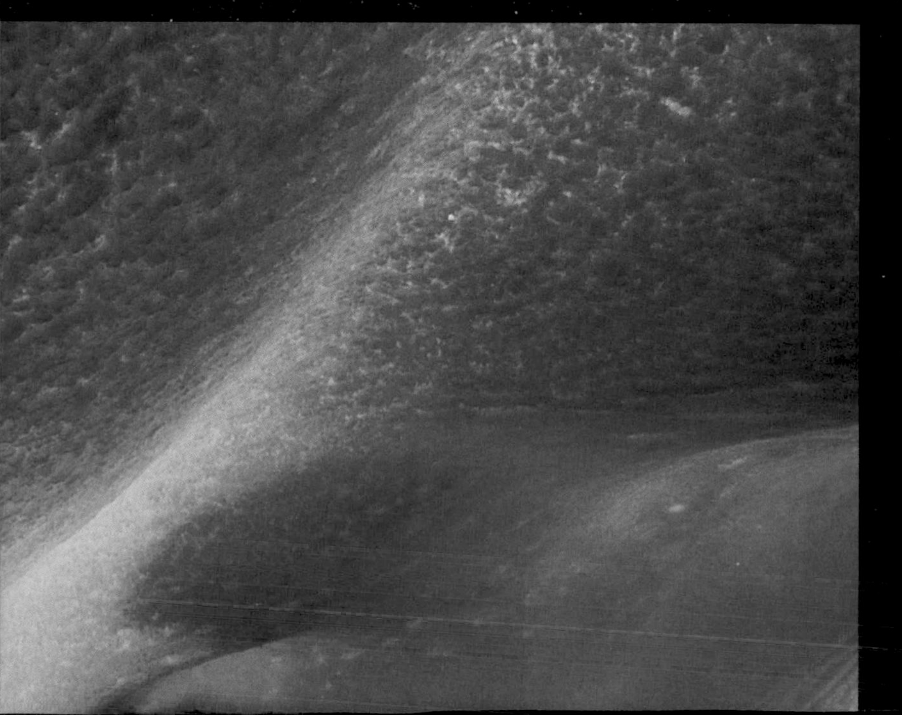

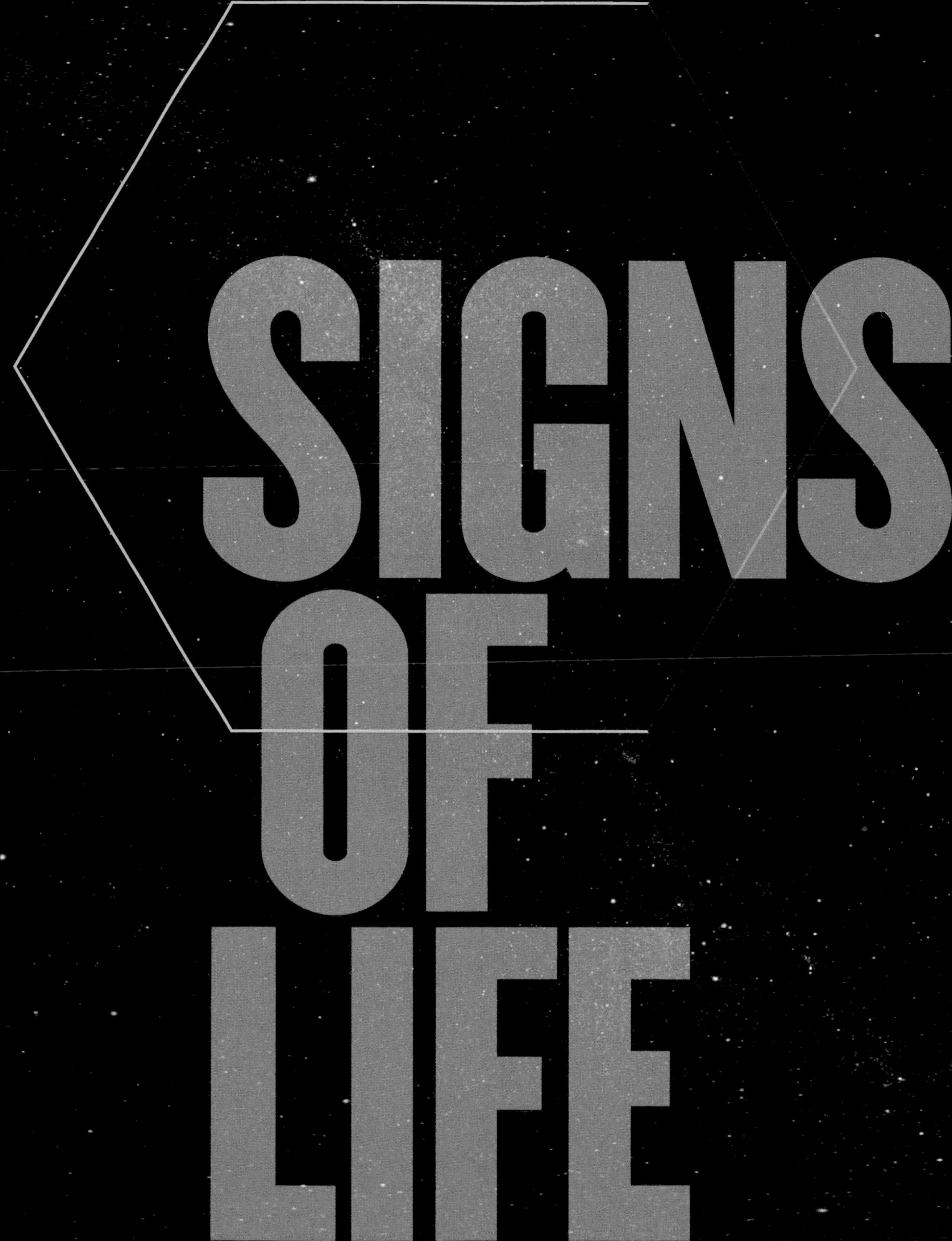
SIGNS
OF
LIFE

Under the blue dome of an ice cave on Antarctica's Mount Erebus—one of the coldest spots on Earth—microbiologist Craig Cary is taking samples, searching for extreme life-forms on Earth that may offer hints about the life to be found on Mars.

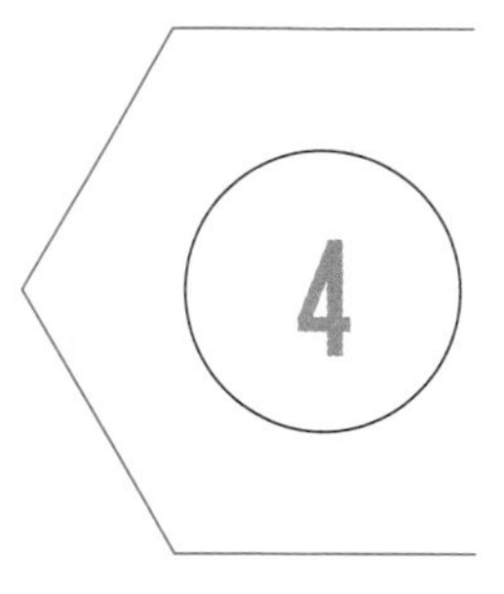

Signs of Life

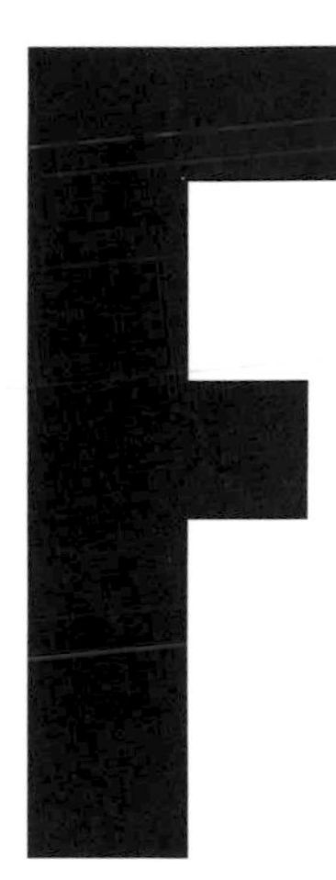

OUR DECADES HAVE NOW PASSED since the touchdown on Mars of America's Viking 1 and Viking 2 landers, sent to probe the prospect of Martian life, extinct or extant. Years of interpreting the data and 26 life-seeking experiments later, the pioneering robots seem to have responded, "Can you repeat the question?" While most researchers involved in the Viking project believe it reported a nondetection of life on Mars, that judgment was not unanimous, and the quest continues. Decades later and billions of dollars in Mars spacecraft costs borne by several nations, the inquiry about past or present life on the planet remains alive and well—even if life on Mars is long gone or never even existed.

Since those first robotic landings on Mars by the Viking missions, "advanced scientific exploration of Mars is in full swing, with ongoing investigations into the history of climate, the possibility of records of past life, and a continuing focus on the habitability of the planet," says James Garvin, chief scientist at NASA's Goddard Space Flight Center and a member of the Mars Science Laboratory/Curiosity Mars rover science team. The Red Planet is one the world's most significant scientific frontiers, he adds, citing recent discoveries of organic molecules, variations in the trace gas methane, and the compelling geological history of the planet involving sedimentary systems and the critical role of water.

What kind of living thing could tolerate the harsh conditions on Mars? Scientists at the German Aerospace Center tested cyanobacteria, a primordial Earth life-form, subjecting it to radiation, low pressure, temperature extremes, and other stresses—and it survived.

The next steps in Mars exploration are now at hand, Garvin says, with increasingly sophisticated missions ahead. "Robotic missions will serve to pave the way for a transition in the 2020s to an era in which preparation for human exploration emerges as NASA continues its journey to Mars into the 2030s."

NASA's next nuclear-powered Mars rover, slated to launch in 2020, will join the Curiosity rover to scan terrain selected for its geological diversity, looking for signs of past life and collecting Mars samples for eventual shipment back to Earth—

EPISODE 4

POWER

Four years have passed since the first crew landed on Mars, and against the odds a settlement has been established. It is a stable if modest outpost of the first humans to inhabit the planet. Plans call for steady expansion of the settlement, and a team of superstar scientists has been sent to make sure that deadlines are met. The time line for expansion is faster than expected, though, and a Martian dust storm moves in over the fragile settlement, threatening to delay progress.

a costly and somewhat controversial task. Whereas the risks to Earth from being on the receiving end of Mars samples are deemed very low, they are not zero. Hauling back Martian samples could well mean dealing with biological hot property, not to mention heated rhetoric and public concern about creepy-crawlies from Mars eating away at Earth's biosphere, something like the catastrophe depicted in Michael Crichton's *Andromeda Strain*.

Explorers will need to be careful even about what they track into their Martian habitations from outside. Consider what Apollo moon walkers encountered: After strutting around on the moon and crawling back into their lunar module, Apollo crew members noticed they had tracked lunar dust particles into their residence. In doffing their helmets, some likened the aroma to wet ashes in a fireplace. "All I can say is that everyone's instant impression of the smell was that of spent gunpowder," recalls Apollo 17 astronaut Harrison "Jack" Schmitt, who walked the moon's surface in December 1972. So for 21st-century astronauts, setting both foot and nose on Mars are factors to weigh when designing live-in habitats—and establishing live-in habits.

The perchlorate problem

ANOTHER CHALLENGE THAT the planet tosses at future explorers is the toxic perchlorates that are pervasive on the Red Planet. This carpet of chemicals could boost the chances that microbial life exists on the Red Planet, but they are also perilous to the health of expeditionary crews. Endocrine-disrupting compounds, these salts are powerful thyroid toxins.

Perchlorates are reactive chemicals first detected in the soil of Mars's north pole by NASA's Phoenix lander in May 2008. More recently, the Curiosity rover detected perchlorates in Gale Crater, the area on Mars where it landed in August 2012. Finding calcium perchlorate was an unexpected result, explains Peter Smith, the Phoenix principal investigator at the University of Arizona in Tucson. "*Perchlorate* is not a common word in the English language. All of us nonchemists had to go and look it up," he admits.

Microbes on Earth use perchlorates as an energy source, Smith says. They actually live off highly oxidized chlorine, and in reducing the chlorine down to chloride, they use the energy in that transaction to power themselves. In fact, when there's too much perchlorate in drinking water, microbes are used to clean it up. What's more, the seasonal flows seen on Mars may be caused by high concentrations of the brines of perchlorate, which has a strong attraction to water and can drastically lower its freezing point.

There are two sides to the discovery of perchlorates on Mars. On the one hand, they are toxic to humans. Mars walkers will find it hard to avoid dust particles sticking to surface equipment, and so perchlorates are likely to find their way into a habitat. Martian dust devils laden with perchlorates are sure to be devilishly dangerous too.

On the other hand, perchlorate is a key chemical component within the pyrotechnics industry, and ammonium perchlorate is a component of solid rocket fuel. So mining perchlorates may turn out to be an on-the-spot resource for fuel to travel on and even depart from the planet. Some researchers advocate a biochemical approach for the removal of perchlorate from Martian soil, one that could obtain oxygen for human consumption and fuel but would also be energetically cheap and environmentally friendly.

The general consensus is that we do need to worry about perchlorates on Mars, but they do not present an insurmountable problem. "The current view of perchlorates/chlorates on Mars is that they are probably globally distributed," says Doug Archer, a scientist with the Astromaterials Research and Exploration Science Directorate of NASA's Johnson Space Center. There are likely areas on the planet that exhibit lower amounts, and perhaps some with higher. It turns out that perchlorate was used for years as a pharmaceutical to treat hyperthyroidism. "So we know quite a bit about human exposure to perchlorate," continues Archer. "And in small amounts, as long as you are taking iodine supplements, there don't seem to be any deleterious effects."

In fact, the veneer of globally distributed perchlorate on Mars could actually aid human exploration, believes Archer. Perchlorate is a very effective desiccant (meaning it has a high affinity for water), and processing it to release its water is technically feasible. In addition, high-heating perchlorate causes it to decompose and release oxygen, a handy necessity for crews on Mars.

Overall, though, perchlorates on Mars are a complicating factor in the search for life on—and the quest to bring life to—the Red Planet.

In the matter of microbes

DISCOVERING MARTIAN LIFE is no easy job, explains John Rummel, a senior scientist at the SETI (Search for Extraterrestrial Intelligence) Institute and a former NASA planetary protection officer. Sequestering Earth life from Mars is a huge task as well. "We currently estimate that as many as 300 million live Earth microbes will be launched to Mars on a robotic spacecraft that *isn't* looking for life . . . and perhaps 30,000 on an entire spacecraft *trying to do* life detection," Rummel emphasizes. Many of those microbes won't survive the flight to Mars, he says, and perhaps 99 percent will be killed off by the harsh Martian ultraviolet radiation in the first day or so after landing.

Some will survive, though—and that's from a *nonhuman* mission. "Contrast that to the microbes carried by a single human explorer—about 30 trillion nonhuman microbial cells—all of which *will* survive the journey to Mars, and in style. Such a long passenger list complicates the task of finding microbes that may already be there," Rummel explains.

Scientists are scripting precautionary steps not to violate any life force on Mars. Will humans on Mars, by their very presence, introduce Earth life into places on the Red Planet where it could potentially prosper, and perhaps taint the very signs of life we are looking for? Conversely, if homegrown Martian microbes are found, could space-suited backpackers face danger?

"I think the question of whether life ever originated on Mars, and perhaps died out, is still the prime mover," says William Hartmann, a senior scientist and longtime Mars researcher at the Planetary Science Institute in Tucson and an investigator on the U.S. Mariner 9 Mars orbiter that mapped the Red Planet for the first time in 1972. Looking for life on Mars, Hartmann suggests, is the next step toward answering the question of whether we terrestrial life-forms are alone in the universe. For Hartmann, the key is to follow the water, and especially the history of water and its role in Mars's climate over millions of years. "It's tough, because the surface of Mars is sterilized by ultraviolet and mostly very dry," Hartmann explains. "So if we are to understand the history, modern role, and presence of water and ice on Mars, we have to deal with subsurface

At Mars's Hale Crater, orbiting cameras have revealed dark streaks (or RSL—recurring slope lineae) suggesting seasonal water flow going on right now. An orbiting spectrometer has detected hydrated salts: liquid salt water coursing down the crater in stretches as long as a football field.

processes. Surely there are massive underground ice deposits, as known since the Viking era," Hartmann explains. A central question may be whether there were always interconnected underground aquifers on Mars where microbes could survive, moving from one geothermal area to another as the planet evolved.

Exploring nether regions

THE MORE WE SEE OF MARS, the more experts think they are finding evidence of water. "Some of the most puzzling and least understood features of Mars are found in regions that, due to topography and surface geology, are not accessible to investigation by rovers or landers," says Joel Levine of the College of William and Mary. Examples of out-of-the-way territory, he notes, include areas of strong crustal magnetism, regions of production and emission of atmospheric methane, and the walls of several craters that suggest the existence of transient, short-lived, flowing water.

Exciting evidence for flowing water on Mars came from work led by Lujendra Ojha of the Georgia Institute of Technology. A group of experts made use of instruments on board the Mars Reconnaissance Orbiter: the high-resolution imaging science experiment camera and the compact reconnaissance imaging spectrometer for Mars. This Mars-orbiting gear cross-haired what are branded as recurring slope lineae, or RSL in Mars shorthand.

“Just think of the adventure of four astronauts catching a sunrise from the surface of Phobos, or the thrill of their view of Mars from above. The whole world would be with them, exploring in spirit on the most daring adventure in human history.”

—Bill Nye, CEO of the Planetary Society

RSL form and snake down steep slopes on the planet during warm seasons when temperatures exceed minus 10°F. They disappear at colder times during the Martian year. The Mars Reconnaissance Orbiter has techniques to identify minerals on the slopes striped with these puzzling RSL. Researchers found distinct mineral signatures in the RSL that were particularly wide in diameter.

And then the clincher came when the researchers looked at the same RSL locations at another time in the Mars year, a time when the RSL weren't visible—and the hydration signatures had disappeared.

Something is hydrating these salts, says Ojha. Furthermore, the visible streaks seem to come and go with the seasons. "This means the water on Mars is briny rather than pure. It makes sense, because salts lower the freezing point of water. Even if RSL are slightly underground, where it's even colder than the surface temperature, the salts would keep the water in a liquid form and allow it to creep down Martian slopes," he suggests.

Researchers now believe that these mineral signatures are caused by perchlorates—a finding that locates perchlorates in entirely different areas from where earlier Mars landers explored. It's the first time perchlorates have been identified from orbit—and, more dramatic, the first observation that unequivocally supports the hypothesis that liquid water exists on Mars, forming the distinctive rivulet patterns characteristic of the RSL.

Where there's water . . .

Resolving whether liquid water exists on the Martian surface is central not only to comprehending the hydrologic cycle on Mars; it is also an essential factor in the search for existing life on the Red Planet. Ojha and his colleagues caution that while there are fleetingly wet conditions near the surface on Mars, the water activity in perchlorate solutions may be too low to support known life—at least forms of life as we know them here on Earth.

James Wray, a Georgia Tech planetary scientist, wants a Mars mission to get up close and personal with these alluring features. "Personally, I think we could learn a lot about RSL from landing near them . . . and then driving close enough to image, but not to touch," he says.

It will be the best way to observe their activity, Wray adds: From Mars orbit, there's no way to watch a given RSL evolve from one hour to the next, day after day. "We could do this easily on the ground, even from a stationary lander," he points out. Soon the scientists' hunger to know more about the chemistry and organic content of RSL will evolve until the questions require true "contact science," perhaps by using a sterilized probe.

Similarly, David Stillman of the Southwest Research Institute in Boulder, Colorado, argues that the search for life within RSL should be a foremost priority. He cautions, however, that RSL may be far too briny for any known life-form to respire there. This factor could minimize the impact of cross-contamination, thus reducing planetary protection concerns. But perhaps Martian life evolved a way to live in such an environment? Or maybe there is life within the depths of the RSL source regions?

In other studies, Stillman and Robert Grimm, also of the Southwest Research Institute, analyzed the seasonal water budget of an RSL site in Mars's huge Valles Marineris canyon system. Their research suggests that the site is recharged by an aquifer.

In fact, the team estimates that the total amount of water liberated from the area is equivalent to 8 to 17 Olympic-size swimming pools. The only way to annually recharge such a large volume of water, Stillman suggests, is via an aquifer. Stillman and his colleagues point to fractures that extend into a regional pressurized aquifer as possible sources for the brine that is reaching the Martian surface. Furthermore, the water lost from the myriad RSL within Valles Marineris—roughly 50 sites—is likely a significant regional source of atmospheric water vapor for much of each year. Maybe there is a zone of salty regolith within a few inches of the surface at these sites.

Mars's recurring slope lineae remain truly fascinating features, and they may well be the best places to search for life. "There are more RSL than ever, 263 sites over a larger geographic area . . . and we still are having a difficult time explaining them," Stillman concludes.

Meanwhile, back on Earth

A HARSH ENVIRONMENT with very little water yet flooded by intense ultraviolet radiation: Mars? No, the Atacama Desert in Chile, a brutal place found to have microbial colonies underground or inside rocks, and a place where scientists are hard at work to learn more about extremophiles, organisms that thrive in the extremes of altitude, cold, darkness, dryness, heat, mineralized environments, pressure, radiation, and vacuum—organisms, in other words, that have a lot to teach about the possibilities of life-forms on Mars.

Recently, more than 20 scientists from the United States, Chile, Spain, and France completed a month of Atacama Rover Astrobiology Drilling Studies (ARADS). One facet of the project, led by the NASA Ames Research Center, collected samples for laboratory investigations of the extreme microorganisms living inside salt habitats in the Atacama. These distinctive, tough-as-nails microorganisms should improve life-detection technology and strategies for implementation on Mars. Over the next

Rock formations from two different regions of Mars indicate different environments. "Wopmay" rock from Endurance Crater (left), studied by the Opportunity rover, suggests an ancient environment with water high in acid and salinity, not conducive to life. On the other hand, "Sheepbed" rocks from Yellowknife Bay (right), visited by the Curiosity rover, show fine sediments that were once underwater in an ancient habitable environment.

four years, the ARADS project will return to the Atacama to demonstrate the feasibility of integrated roving, drilling, and life-detection techniques for finding evidence of life on Mars.

Until humans set foot on the Red Planet, ARADS can help hone the expertise to identify a location of high probability for current or ancient life on Mars, place a drill, and control the operation robotically.

The value of water

WHILE HAVING LOADS OF WATER on Mars does increase the potential for thriving Martian organisms, available water on Mars is also a driving force behind a persistent human presence on the planet. Are these two notions reconcilable or conflicting situations?

Some hypothesize that life-forms on Mars and Earth are distant, ancient cousins. That outlook is the panspermia theory, which holds that microbial flotsam from Mars traveled by meteorite and seeded life here on Earth, an upshot of the rough-and-tumble process of planetary formation. If so, in essence, we are all Martians. Right or wrong—or something in between—by making our way to Mars, the human race could, in truth, be completing a biological round-trip.

But more likely, even if they were the same hundreds of millions of years ago, life-forms on Mars and Earth are now very different. We need to keep it that way, argue many. "Earth life needs water to grow and, very likely, so does Mars life . . . if it exists," says Catharine Conley, NASA's current planetary protection officer. "This means it's very important to understand the potential for Mars life to be present in any water sources we access on Mars. This is to protect astronaut health, among other things. We don't know if Mars life could be hazardous to us, so being careful until we know more makes sense." Something that is known, she continues, is that introducing Earth life to Mars could interfere with future human interests at the planet.

"People boil water when camping, because water sources on Earth are often unhealthy to drink because of microbial contamination. Contaminated aquifers on Mars would impose an extra burden in energy and hardware to mitigate. That would increase the cost and risk of human activities," Conley says. "Fortunately, we know how to kill Earth organisms. So the solution is pretty straightforward. Sterilize the surfaces of hardware that's accessing locations where water could host Earth life."

"The focus on protecting Mars, from the beginning of space exploration, has been that we should do the careful search for Mars life early on, and then decide what to do later on the basis of the best scientific information we've collected so far," says Conley. Looking back to the Viking spacecraft days, she points out that the landers were carefully cleaned and sterilized before takeoff in an effort to take special care to avoid introducing Earth life before researchers understood the Mars environment.

"Also the Viking project team understood it would be trivial to find life on Mars . . . just bring it with them! The challenge was finding Mars life," Conley says, and certainly not finding microbes from Earth brought along for the ride. "After Viking data seemed to indicate that Mars was cold, dry, and dead, we relaxed the requirements for Mars, to allow a half-million heat-resistant microbes per spacecraft to land on Mars."

DISASTERS

Hostile life-forms | Unknown microbes may exist on Mars in a suspended state. Brought to life by water or heat we introduce, they could invade our bodies. Even if they don't threaten our health, they could gum up our life support machinery.

What could go wrong?

More recently, however, ever since Mars orbiters observed possible water flows, the industry has been more careful about contaminating Mars with Earth microbes. The internationally recognized concept of protected special regions has been put into place, and spacecraft destined for these locations must be cleaned to Viking standards. "What Viking did 40 years ago is the maximum level we would protect anywhere on Mars today," Conley explains. The Viking spacecraft was sterilized with heat, but there are other established methods that Conley points to, such as vapor hydrogen peroxide, gas plasma, and various types of radiation.

"Novel approaches like antimicrobial coatings could also help," she adds. "It is an important area for technology development. If someone comes up with a clever solution, it could also improve life on Earth for people who currently don't have access to clean water."

Going green

LONG-TERM PLANS for keeping Mars crews happy and healthy must include growing plants, recycling air and water within a habitat, and providing a source of fresh food. A method for growing crops that will sustain a crew and not just supplement meals made on Earth "still has some distance to go," reports a study led by Scott Smith, a NASA senior nutritionist and manager for nutritional biochemistry.

Astronauts are already harvesting food in space. Of late, crew members on the International Space Station have grown a variety of red romaine lettuce in a deployable plant growth system that provides light and nutrients but depends on the space station's environment for temperature control and carbon dioxide.

Ray Wheeler, lead for advanced life support research activities in the Kennedy Space Center's Exploration Research and Technology Program, says that once space-grown salad crops are perfected, potatoes, wheat, and soybeans are next—items that, along with salad greens, may provide a more well-rounded diet.

The water that has recently been discovered on Mars does not necessarily make growing food any easier, says Rob Mueller, senior technologist in the same Exploration Research and Technology Program at Kennedy. Drawn from RSL, Mars water will be briny and will need treatment to remove perchlorates and other impurities. The Red Planet receives only 43 percent of the sunlight that Earth receives, and some regions of the planet will never receive adequate light to grow plants. Any greenhouse will have to be able to protect the plants growing inside from intense radiation and extreme temperature swings.

Given these challenges, Ray Wheeler suggests one future scenario: We may be able to transport water, pumps, and fertilizer salts to Mars and grow plants hydroponically, inside a protected environment. Use high-intensity LED lights "to help push the plants hard," he recommends. In the long run, Martian soils might be treated and used as the growing systems expand.

Plant ecologist Wieger Wamelink of Wageningen UR, a university and research center in the Netherlands, believes that the Martian soil can grow food for visiting humans. In a recent pilot experiment, he cultivated 14 plant varieties in mock Mars soil made of the volcanic soil of Hawaii. To Wamelink's surprise, the plants grew well. Some even blossomed. "I had expected the germination process to work, but I thought the plants would die due to a lack of nutrients," he says. Soil analyses show that Mars soil contains more nutrients than expected: not only phosphorus and iron oxides but also nitrogen, an essential plant nutrient

Over time, the Martian standard of living for humans will evolve. Mars is a world of surprises. What we do know, though, is that there will come a day when "life on Mars" includes human beings and the plants they grow. ■

CAVE OF CRYSTALS

Discovered only in 2000, this amazing cave deep within Naica, a mountain in Mexico's Chihuahuan Desert, supports extreme forms of virus and bacteria, Earth-dwelling life-forms that suggest what might be found on Mars. Within these columns of selenite—a crystalline variety of gypsum—the humidity is 90 percent and the temperature can reach 118°F.

AT THE CORE

Drilling into an ice-tower wall at Antarctica's Mount Erebus, researchers hope the ice core contains archaea or other microbes lofted from deep within a volcano and frozen in the tower ice—life-forms that may presage what we will find on Mars.

Cave Explorer

HEROES | PENELOPE BOSTON

Director, NASA Astrobiology Institute (NAI)

The odds of locating life on Mars may go up by going down—plunging into deep underground caves, according to Penny Boston, an astrobiologist and speleologist. Given the planet's environment, the prospects of finding signs of life topside seem decidedly dreary: It's extremely cold. The atmosphere is thin. The reddish surface is highly corrosive and oxidizing, and the planet is blitzed by galactic cosmic rays and solar storms. A mix of the wrong stuff.

Have heart, says Boston, a notable geologist who's keen on exploring caves. "If we just want to go dabble around the surface of Mars, we're never going to find life," she says. "The circumstances are too difficult . . . and have been too difficult for too long. It's certainly inhibitory to microbes." Potential exists if we get into natural cavities on the planet, Boston says—features that could have preserved a biological and climate record from early Mars or even more recently. "The subsurface is great at preservation," she says. "It's reasonable to expect on Mars that there will be significantly fewer weathering effects on materials than the surface materials we're so obsessed with now."

Clinging robots that can walk up walls and crawl on cave ceilings are in the design and testing stages. And there's thought that lava tubes on Mars could become natural real estate, ideal for conversion into "human-usable" spaces. "I'm a big advocate of eventual human exploration of Mars," the astrobiologist explains. "There are ways to deal with planetary protection issues, such as the idea of zones. In our cave work, we have sacrifice zones. That's where you have human operations but contain and control the amount of contamination by having different protocols."

Are we ready to do that now? "No, and we need to develop those approaches and test them here on Earth in a more rigorous way," Boston says. She is of the mind that a protracted human presence on Mars is warranted. "I do think becoming a multiplanet species is part of our destiny. If we don't proceed, then we are a single-planet species until something terrible happens, and then we're done." In the long term, the work of terraforming Mars—adjusting the surface and atmosphere of the planet to support human life—"is going to be done by the people who are living on Mars," Boston predicts. "You are not going to figure it out here on Earth and then ship terraforming kits to those on Mars. It's likely going to be a natural outgrowth of humans who go there and chose to continue to live there. Ethically and politically, terraforming should be a decision made by the people who have the most skin in the game—by those who are permanently there."

From the walls of Mexico's eerie Cueva de Villa Luz ("cave of the lighted house"), cave expert Penelope Boston coaxes a glob of ooze lovingly nicknamed a snottite—a cluster of microbes that thrive on hydrogen sulfide in an environment toxic to humans and most other Earthbound life-forms.

HIDDEN LIFE-FORMS

Extremophiles survive on Earth in the most unlikely places, such as half a mile beneath the Antarctic ice, where this microbe was retrieved. Might life-forms on Mars have retreated from the planet's surface to ice caves below in ages gone by?

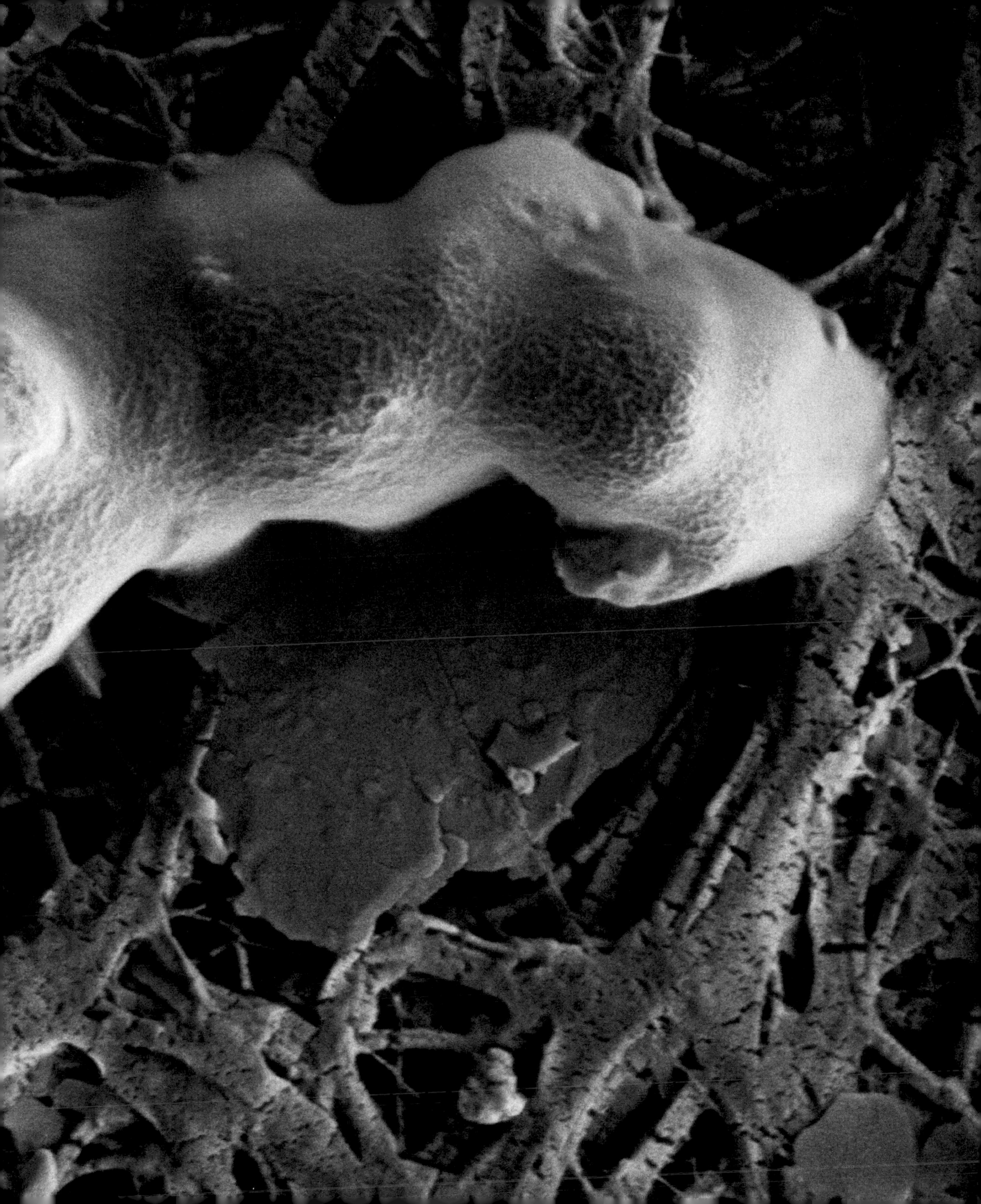

CLEAN SWEEP

Schiaparelli, the entry-descent-landing module of the European Space Agency's ExoMars mission, gets a final scrub before its March 2016 launch. Inspection samples taken on Earth before launch are part of a strict planetary protection protocol that all Mars-bound vehicles must undergo.

LIFE AT EITHER EXTREME

Perhaps lichens—a tough life-form that comes from the union of a fungus plus algae or cyanobacteria—could exist on Mars. This sulfur-yellow one thrives in polar extremes and made it through a Mars simulation experiment. Thermophiles—heat-loving microbes— present in Yellowstone National Park's famous Grand Prismatic Spring (opposite) may also hold clues to Martian life-forms.

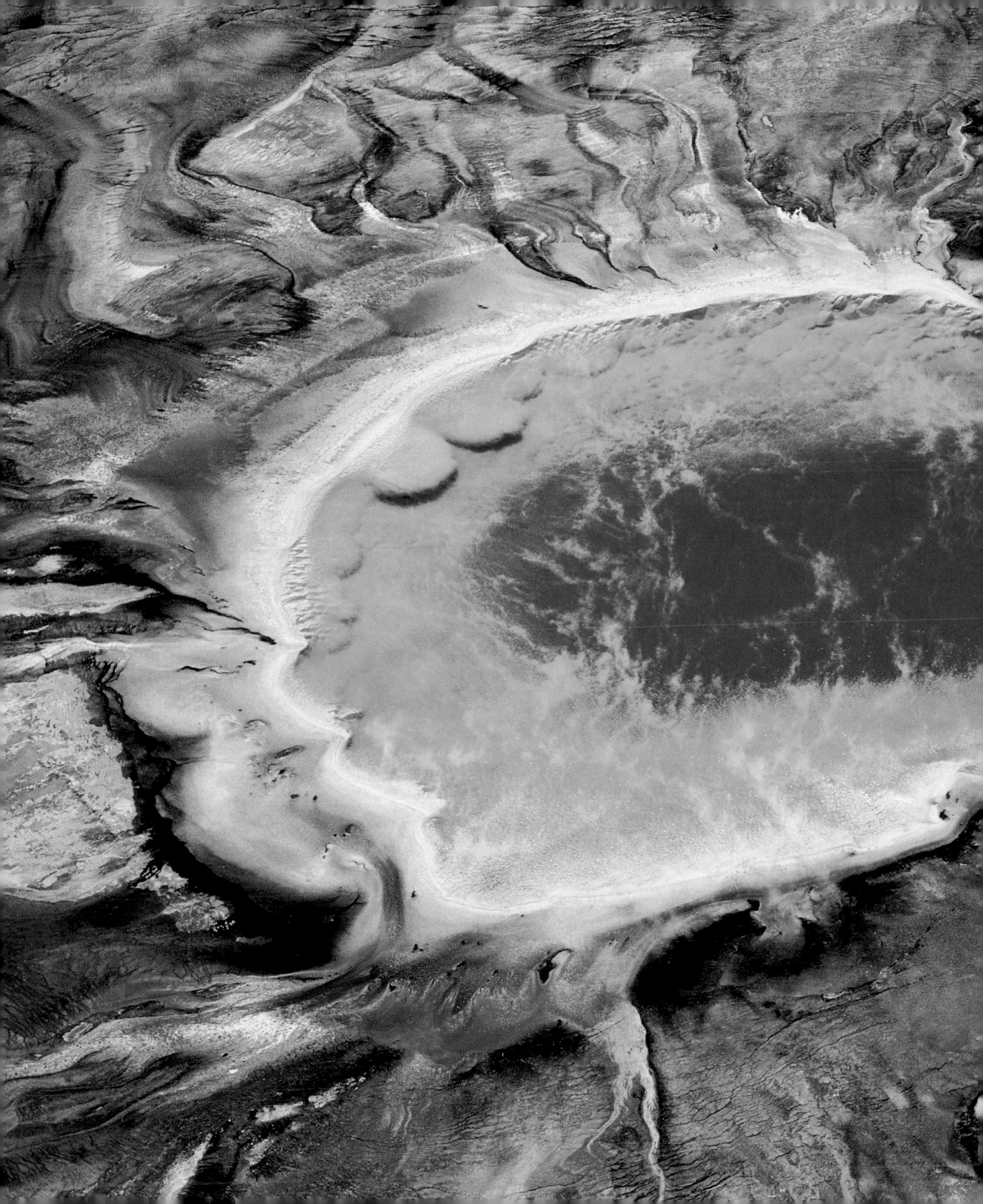

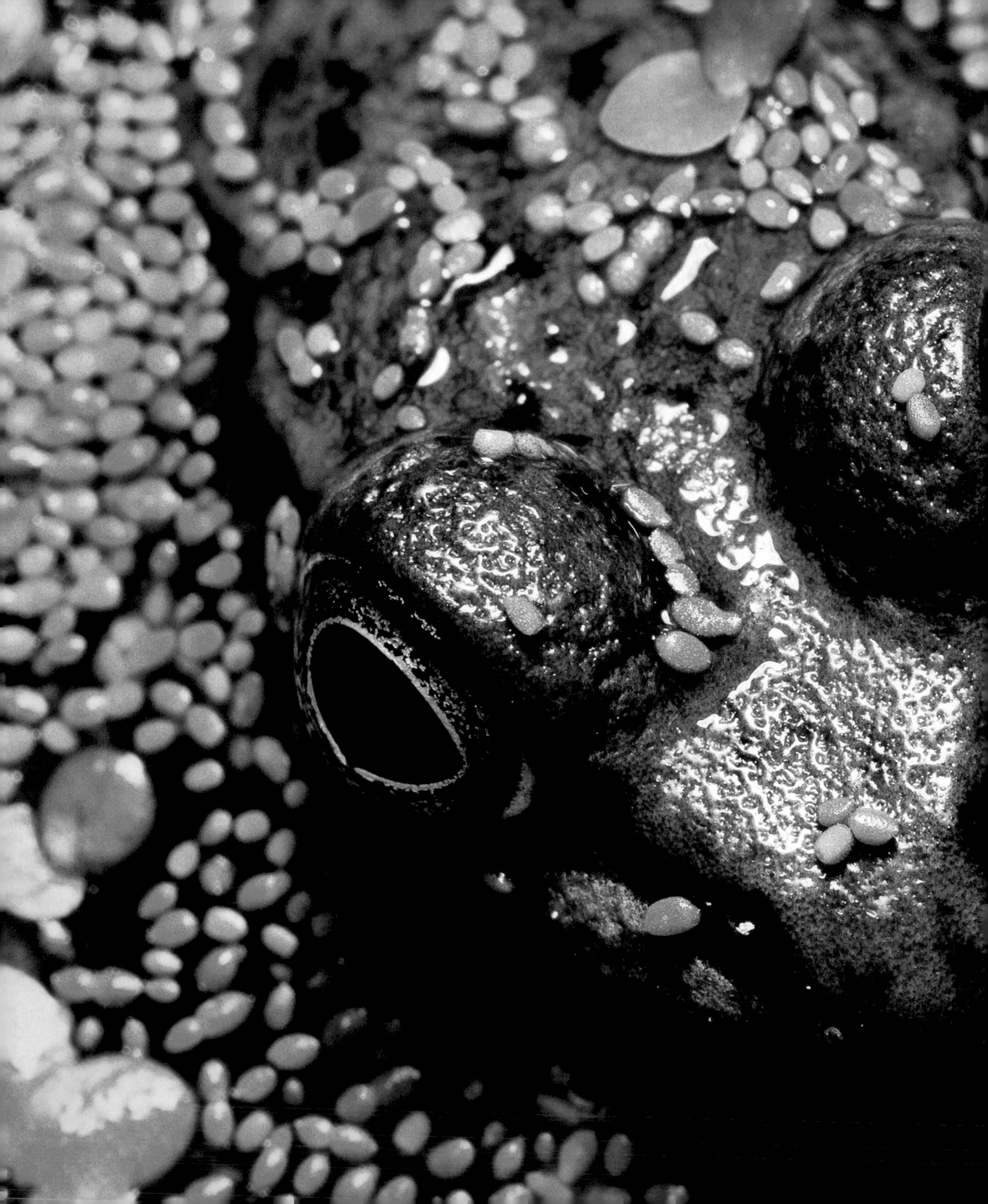

PUSHING THE DEFINITION OF LIFE

The wood frog (*Lithobates sylvaticus*, left) lives through repeated freezes, its internal organs literally stopped cold. The tardigrade (below), an eight-legged microanimal, displays cryptobiosis: That is, it can dry up or freeze in response to extremes of heat, cold, atmospheric pressure and radiation levels. As we learn more about such creatures on Earth, we may be building up knowledge about life on Mars as well.

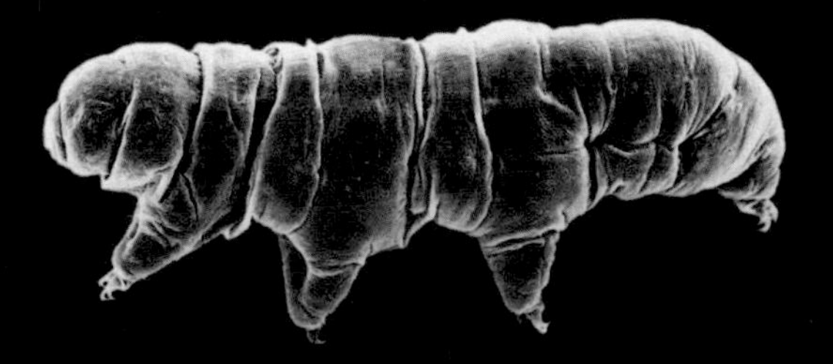

ICE OUTSIDE, LIQUID WITHIN

Jupiter's moon Europa reminds us that a frozen exterior may be hiding liquid water within. Observations in recent decades show that Europa is an icy shell surrounding a deep ocean. Spacecraft have provided observations of water pluming up and chunks of ice breaking off and re-forming, hence the weblike pattern on Europa's surface.

PRACTICING SELF-SUFFICIENCY

In preparation for the Mars One enterprise, researchers used simulated Martian soil and successfully grew tomatoes (below). Meanwhile, within similar restrictions, homesteaders living at the Mars Desert Research Station in Utah grew Swiss chard (right). All such sample vegetable gardens are closely monitored to see how much light, water, and soil nutrients plants will need to thrive inside a Martian habitation.

Halifax 2008
www.conjugatemargins.com

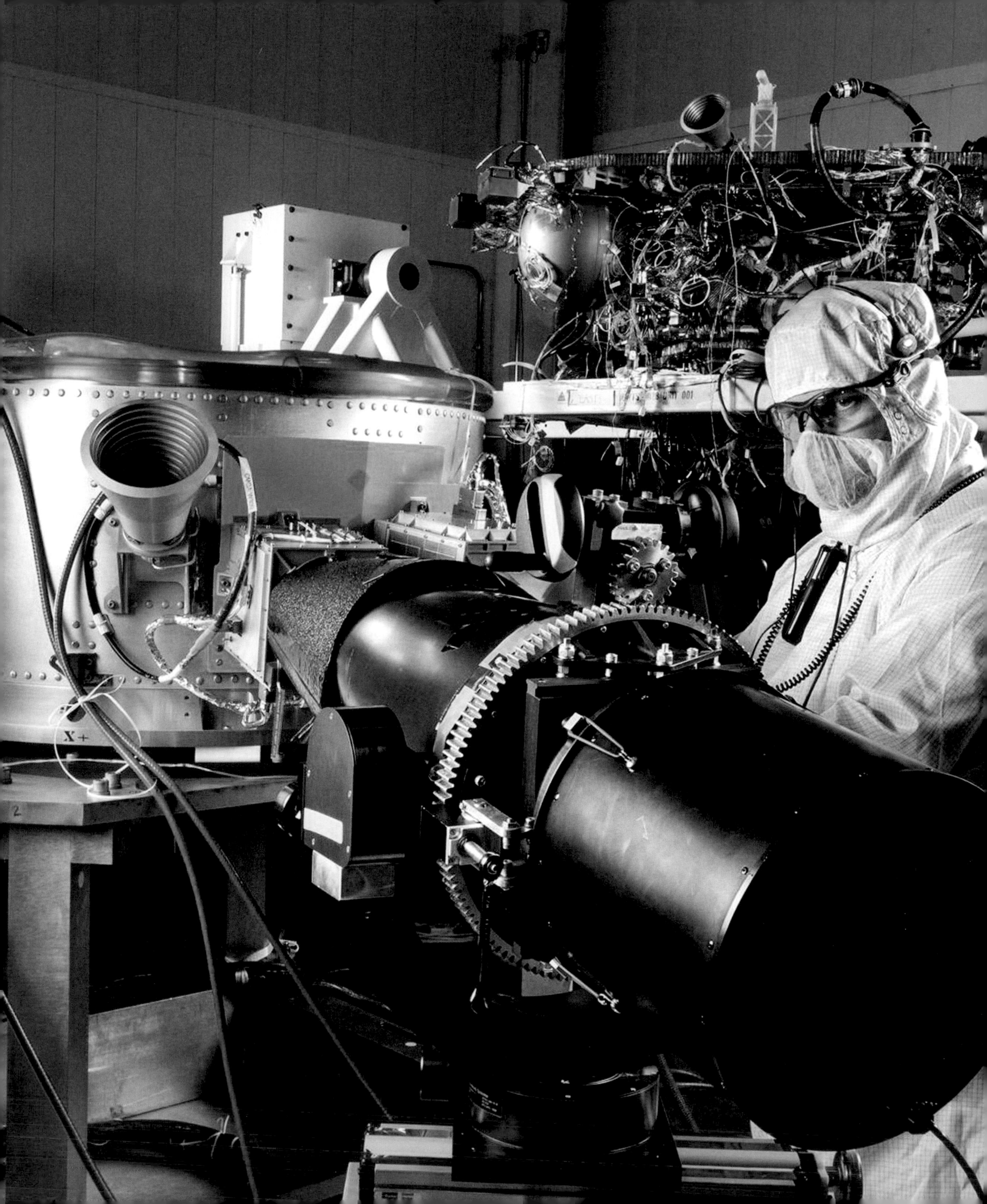
X+

HEROES | CATHARINE "CASSIE" CONLEY

Planetary Protection Officer, Office of Planetary Protection, NASA

Just how tough is it to be NASA's planetary protection officer? "It's a little bit like being a police officer," Cassie Conley responds, "or a kindergarten teacher."

Most people are perfectly happy to follow the international consensus rules, Conley says, because they understand that the rules make sense and will protect everybody into the future. "However, there are always a few people who don't want to follow them, for whatever reason. It's like the guy in your college dorm who drinks out of the milk carton."

Responsible exploration of the solar system means protecting the science, the environments that are explored, and the Earth. And the credo of the Office of Planetary Protection is, "All of the planets, all of the time." That's a tall order, and the aims are numerous; among them are preserving our ability to study other worlds as they exist in their natural states; avoiding the biological contamination of explored environments that may obscure our ability to find life elsewhere, if it exists; and ensuring prudent precautions to protect Earth's biosphere in case life does exist elsewhere. Ultimately the objective is to support the scientific study of chemical evolution and the origins of life in the solar system. "We know, from invasive species on Earth, that once life is introduced, it's very hard to get rid of it again," says Conley. It's frustrating, she adds, when people don't follow the rules because it's very easy for one person's or a project's actions to cause problems for everyone else.

Conley's work addresses many facets of mission development, including assistance in the construction of sterile—or low-biological-burden—spacecraft. She's also engaged in the shaping of flight plans that protect planetary bodies of interest. Additionally, there's the task of guarding the Earth from returned extraterrestrial samples.

Mars exploration requires a phased approach. "Be careful early," Conley cautions. "Tailor later constraints using knowledge gained." Humans have many interests in Mars, and understanding potential hazards supports all of them. Searching for Mars life becomes more difficult the more Earth contamination is introduced. Future colonization of the Red Planet could be challenged if unwanted Earth-invasive species are introduced, Conley notes. "If you're going somewhere to look for life, don't trash the place or samples before you have a chance to find it!"

Conley points out that in 2003, being a planetary protection officer was voted #17 of the "Worst Jobs in Science" by *Popular Science* magazine—but somebody's got to do it. "But if anything goes wrong with samples brought back to Earth," says Conley, "the world will blame the planetary protection officer."

At Lockheed Martin Space Systems in Denver, a technician inspects key components of InSight (Interior Exploration using Seismic Investigations, Geodesy and Heat Transport), destined to study the deep geology of Mars. Multiple efforts are made to secure the cleanliness and safety of each spacecraft.

MARS IN A BOTTLE

To discover the Earth life-forms that might presage what could be found on Mars, scientists at the German Aerospace Center have devised a chamber that mimics the extremes of a Martian environment: ultraviolet radiation, infrared radiation, soil constituents, low atmospheric pressure, the gas mix of the Mars atmosphere, and temperatures ranging from minus 50°F and lower up to 70°F.

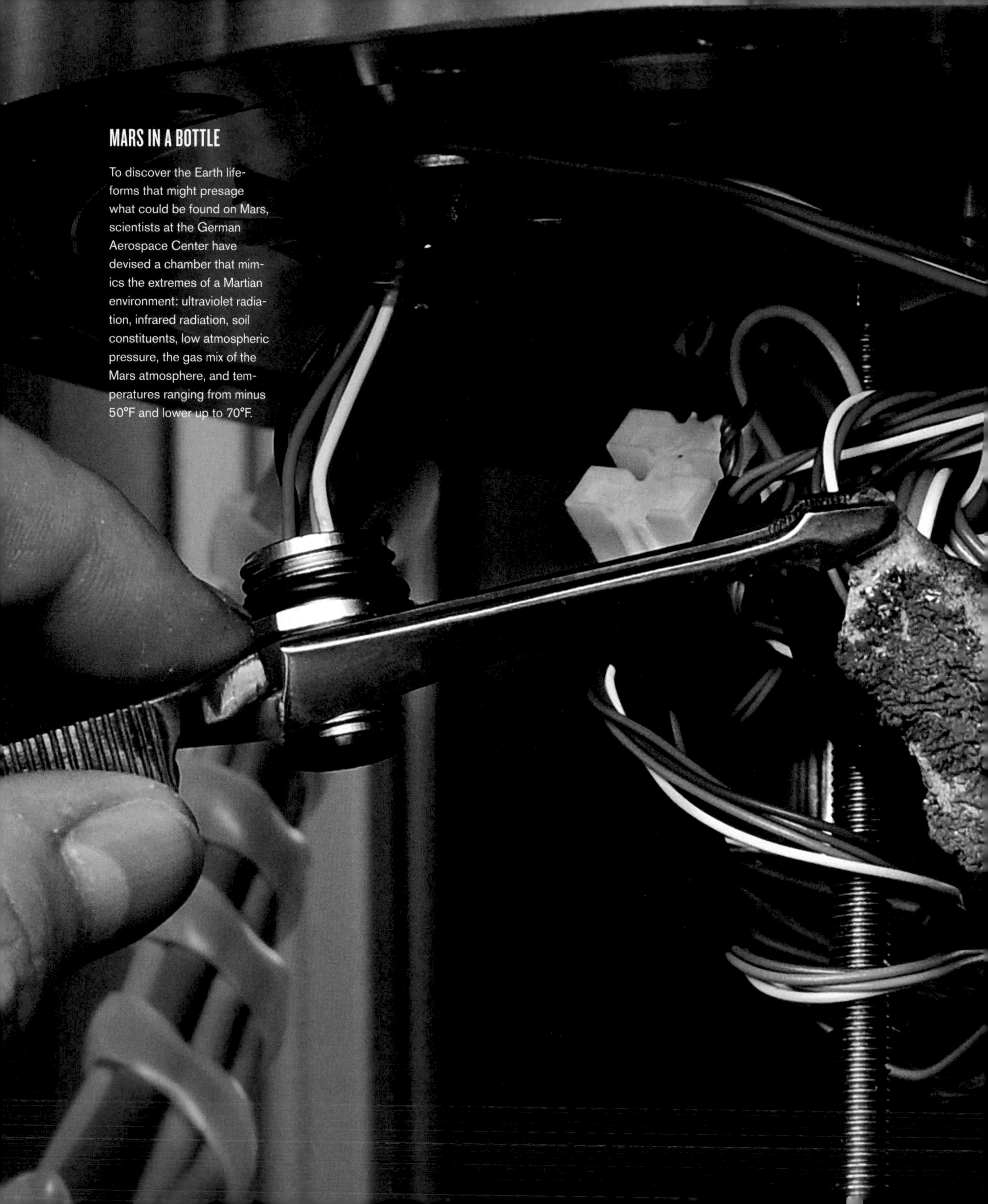

EUROPEAN TOUCHDOWN

The European Space Agency's ExoMars 2020 mission will send this rover to a site on Mars chosen for its likelihood to reveal well-preserved organic material that might illuminate the ancient history of the planet.

DRILLING FOR LIFE

The search for life on Mars proceeds through repeated drilling to analyze the chemistry of the planet. Opportunity's instruments abraded rock and discovered brownish red hematite (below, left), while Curiosity's drill pulled up bluish gray tailings, likely magnetite (below, right), more compatible with life. Curiosity's instrumentation includes a chemistry lab in which drilling samples (right) are analyzed extensively.

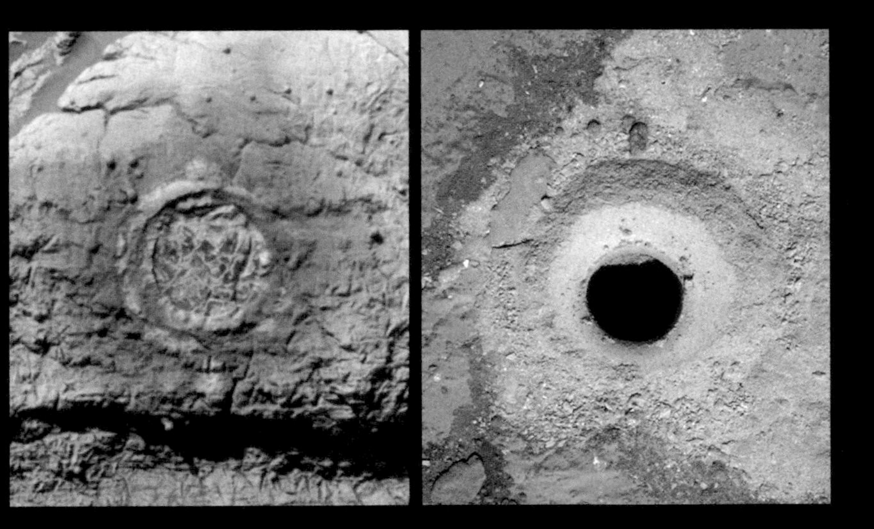

HEROES | CHRIS MCKAY

Planetary Scientist, Space Science Division, NASA's Ames Research Center

It has been 40 years since the first U.S. Mars landers—Viking 1 and Viking 2—gently touched down on the enigmatic world of the Red Planet. Those pioneering robotic missions of the 1970s were geared to probe one question: Is there life on Mars? Now, more than four decades later, that question remains more urgent than ever.

For planetary scientist Christopher McKay, finding the answer has been a long-time quest. McKay has literally gone to the ends of the Earth in his Mars investigations. He has tromped across the Antarctic dry valleys, Siberia, the Canadian Arctic, and the Atacama, Namib, and Sahara deserts—all travels designed to study life in Mars-like environments. McKay's bottom-line mantra, given his research, is, "Drill, baby, drill," adding, "If not drilling down, in my opinion, you might as well not go."

So where on Mars should we go to search for life? McKay's list is short: "All the places that I want to go are underground." The first of three locations he favors is the May 25, 2008, landing site of NASA's Phoenix lander on the low northern plains on Mars, where ice is known to be very close to the surface. "Drilling there, down a meter," he says, "you get to stuff that might have melted a few million years ago."

The second spot is where the Curiosity Mars rover previously explored. McKay says the area didn't get the attention that it deserved. "Yellowknife Bay. At two drill sites, we drilled down two centimeters. We got through mud stone and we reached gray Mars—below the red covering on the surface," he explains. "As far as we can tell, this is sediment that piled up in a bottom of a lake 3.5 billion years ago. We need to get well below the surface so that we're seeing stuff that's shielded from radiation . . . say five meters."

Mars's ancient highlands are third on McKay's list, a location with very strong magnetic fields. "Places that have magnetic fields are very, very old . . . older than anything else on Mars that we see, and they've been relatively undisturbed. You'd need to drill very deep in that terrain, say 100 meters."

As for the value of humans on Mars versus robots, McKay's hands-down favorite is flesh-and-bone explorers. "We have brains, we have eyes, we have feet, and we have hands. Of all those capabilities, the one that's proving the most difficult to extend to Mars remotely and robotically is the hand . . . We need hands to collect rocks and hands to run a drill. These are things you completely take for granted when you are a human scientist in the field."

Chris McKay peers down a hole melted through the surface of Antarctica's permanently ice-covered Lake Hoare, into which he and fellow researchers deploy equipment to observe phenomena that may add to our understanding of the early solar system and the origins of life.

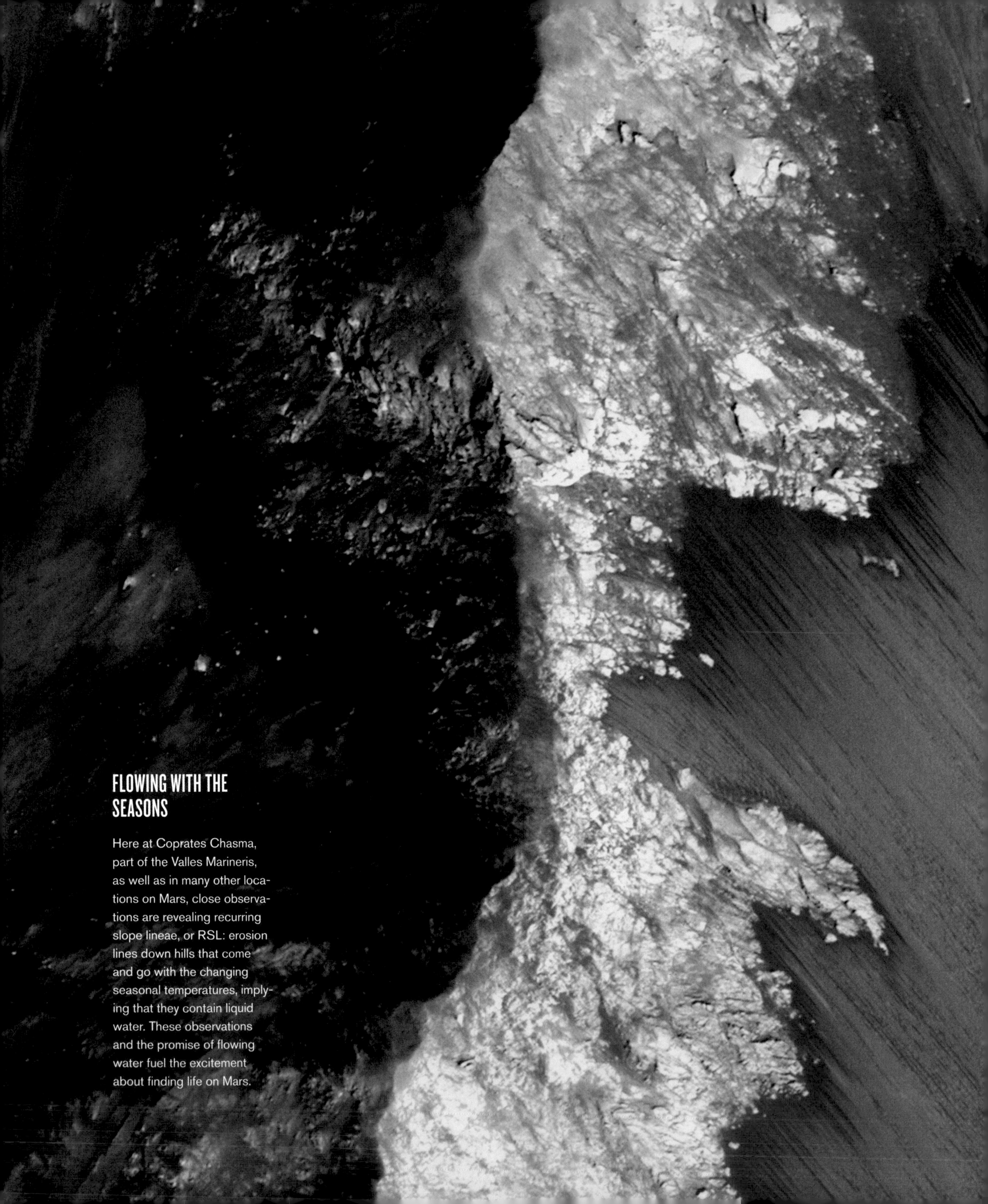

FLOWING WITH THE SEASONS

Here at Coprates Chasma, part of the Valles Marineris, as well as in many other locations on Mars, close observations are revealing recurring slope lineae, or RSL: erosion lines down hills that come and go with the changing seasonal temperatures, implying that they contain liquid water. These observations and the promise of flowing water fuel the excitement about finding life on Mars.

Reaching Mars, we become an interplanetary species. Will it become a place where we transcend national differences or an arena where competition intensifies?

GLOBAL VISION

Earthbound staff at the European Space Agency Control Center in Darmstadt, Germany, track progress during the successful launch of ExoMars 2016, a joint ESA-Roscosmos venture that sent a trace gas orbiter and a lander named Schiaparelli to Mars.

esa
esoc

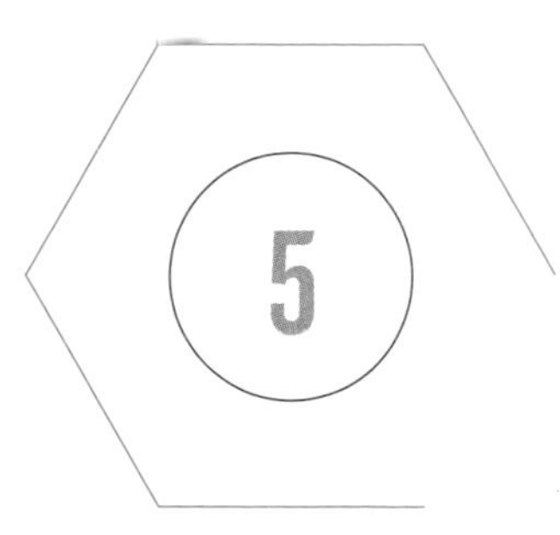

Global Vision

THERE IS A RISING GLOBAL APPETITE to reach for the stunning sights that Mars has to offer. The space race of the early 1960s between the United States and the former Soviet Union is old news. Instead of that 20th-century spirit of one-upmanship, now there's a cadre of countries learning to work together to develop the technological savvy needed to reach Mars. Nations around the world, including some beyond America's customary space partnership base, are thinking about, talking about, and working on sending spacecraft, and eventually humans, to Mars.

Collaborations among Europe, Russia, China, and India, for instance, along with the United States and other spacefaring nations, are likely to result in joined forces that could make the reach for Mars financially feasible and technologically realistic. Equally exciting, growing zeal in the entrepreneurial private sector makes the push to Marsland even stronger and more likely to happen soon.

Today there's a full roster of nations training their sights on Mars:

China: Chinese space officials have indicated that plans are under way to dispatch a rover to the Red Planet as early as 2020. A small-scale prototype Mars rover has been displayed, with indications that the Chinese Mars mission would also collect samples of rocks and soil for return to Earth around 2030. China already has blue-printed and is carrying out a step-by-step robotic lunar exploration venture that is likely to lead to human exploration of the moon. The country is also fabricating the powerful Long March 5 rocket to support a variety of deep space missions.

Europe: An aggressive Mars plan, ExoMars, is under way by the European Space Agency (ESA). The program was kicked off with the March 2016 launch of the trace gas orbiter and an entry, descent, and landing demonstrator module called Schiaparelli, both arriving at the Red Planet in October 2016. The ExoMars initiative also includes a stylish rover, slated for launch in 2020. The undertaking is demonstrating

During a visit to Cape Canaveral in August 2010, President Barack Obama toured a SpaceX launch pad with CEO Elon Musk and gave a major space policy speech, commending "the men and women who work in Florida's aerospace industry" as "some of the most talented and highly trained in the nation."

EPISODE 5

DARKEST DAYS

The dust storm has now lasted for months, and the settlement's infrastructure is suffering as much as its residents' psychological well-being. When the town's power supply is compromised, the lives of its inhabitants are put in grave danger. Undertaking emergency repairs, the team manages to carry on, but the storm confirms that Mars poses both physical and psychological dangers for its early inhabitants.

new technologies that pave the way for a future Mars sample return mission in the 2020s. Both ExoMars missions are being carried out in cooperation between European and Russian space organizations.

India: India's Mars Orbiter Mission, named Mangalyaan, swung into orbit around the planet in September 2014 and marked India's first foray into interplanetary space. The spacecraft is studying Mars features and the planet's atmosphere and is instrumented to scout for the presence of methane that could offer clues to the presence of life. Its success has bolstered the Indian Space Research Organization (ISRO) to contemplate other interplanetary flight. The NASA-ISRO Mars Working Group is building cooperation between the two countries.

Japan: The Japan Aerospace Exploration Agency is considering a go-ahead for a mission involving one of Mars's two moons, Phobos or Deimos, with a landing targeted for the early 2020s and material to be brought back to Earth for thorough analysis. Japan's first Mars explorer, Nozomi (Planet-B), was sent to orbit Mars, but it failed in that mission in December 2003. It is now an artificial planet, flying forever in orbit around the sun.

United Arab Emirates: In the Islamic world's entrance into space exploration, the UAE envisions a Mars orbiter designed to search for connections between today's weather and the ancient climate of the Red Planet. Slated to reach Mars by 2021, the probe will create the first global picture of how the Martian atmosphere changes over the course of a day and seasons. The UAE Space Agency also recently announced a competition for UAE residents to design a habitat for two on Mars. Rules allow that the habitat has to be built with materials that can either be transported from Earth or found on Mars.

Mars on the horizon

When President Barack Obama delivered his 2015 State of the Union address, astronaut Scott Kelly was in the audience, about to embark on his year in outer space. President Obama and Congress acknowledged him with cheers, and then the president urged congressional support for a "reenergized space program," one designed with the goal of "pushing out into the solar system not just to visit, but to stay." Those comments reconfirmed his 2010 address specifically on space exploration, delivered at Florida's Kennedy Space Center on April 15, 2010. "By 2025, we expect new spacecraft designed for long journeys to allow us to begin the first-ever crewed missions beyond the Moon into deep space," President Obama said. "By the mid-2030s, I believe we can send humans to orbit Mars and return them safely to Earth. And a landing on Mars will follow. And I expect to be around to see it."

All that rocket rhetoric aside, there has been White House support for human missions to Mars over numerous presidencies. Most have promoted Mars as the horizon goal of the American space program. Still missing in action, though, is an unswerving and sustainable humans-on-Mars program that is strong politically and financially.

"Mars is the next great frontier for exploration. Exploration is part of the human enterprise, a compelling need. Robotic spaceflight to Mars is a present reality, a precursor to a human journey," says W. Henry Lambright, a professor of public administration, international affairs, and political science at the Maxwell School of Citizenship and Public Affairs at Syracuse University. He is the author of the insightful book *Why Mars: NASA and the Politics of Space Exploration* (2014). "The challenge is to convert a desire into a reality, and that takes political will and expensive hardware. To send humans to Mars will require years of commitment," Lambright says. "America and NASA need to lead an international coalition of spacefaring nations to the Red Planet."

This will be a long-term, arduous quest, Lambright believes, "but if nations share the work and the costs, they can reach the goal in a few decades. There is a need for leadership in this international endeavor, and the United States should provide that leadership. By reaching into space, nations might better learn to work more cooperatively on Earth."

Casting a political eye on the situation is Chris Carberry, CEO of Explore Mars.

Founded in 2010, this private organization, based in Beverly, Massachusetts, is an influential advocate for the goal of landing humans on Mars. "While I wouldn't call current support a slam dunk, I think we have stronger support for Mars in the U.S. Congress and elsewhere than ever before. The question is, How deep is that support, and will it remain if the next administration tries to shift the space program in a different direction?"

To those who may say there is no mandate to go to Mars, Carberry counters with, "Mars has been a horizon goal under several administrations—including the current one—and acknowledged in NASA Authorizations and heavily touted by NASA as its goal, and clearly a topic that fascinates the general public," he observes. "Even if this doesn't qualify as a mandate, it's hard to think of another goal in space exploration that is anywhere as close to achieving that mandate."

All Mars mission advocates agree, however, that it will take a coalition of nations with determination and money. A proven model of space-based international cooperation can be found in the International Space Station, which is expected to stay in operation until at least 2024. The space station is seen as a valuable asset to help mitigate a number of human health risks anticipated on long-duration missions, as well as test and mature the technologies and spacecraft systems necessary for humans to operate safely and productively in deep space.

The European Space Agency is now progressing on the critical path of developing human space transportation systems for missions beyond low Earth orbit with the European Service Module, a key component to be used in conjunction with the NASA-developed Orion crew module for deep space exploration. The new era of coordinated human and robotic space missions requires broad international cooperation, says Thomas Reiter, a German astronaut and now director of the agency's human space flight and operations. Reiter has spent more than 350 days in space, 179 days of those on board the Russian Mir space station. The space station program has demonstrated the importance of a robust international partnership. "Now is the time to build on this partnership and open it to new partners to continue the journey beyond low Earth orbit," he says.

It takes a moon village

MANY IN THE GLOBAL SPACE COMMUNITY see the moon, not Mars, as the next best target. ESA's leader, Johann-Dietrich Wörner, has let it be known he is eager to see a moon village as the next step beyond the International Space Station. He calls Mars a "nice destination" but saves his enthusiasm for a lunar base, global in character with different participants contributing their respective competencies and interests: an ISS for the moon.

Chang'e-3, China's moon mission, reached its destination in late 2013 and has sent back many pictures, including this panorama with Earth in view. A prototype Mars rover, displayed at a 2014 air show, looks similar and is scheduled for a 2020 launch.

But Scott Hubbard, editor in chief of the *New Space* journal, questions whether it is economically possible for NASA to include the moon on its journey to Mars. "I believe that the nation can afford one robust human spaceflight program, but not two," he concludes. The Apollo program, which put men on the moon in the 1960s and 1970s, consumed $150 billion in today's dollars, peaking at 4 percent of the U.S. federal budget in those days, says Hubbard. Yes, Apollo changed the course of human history, "but that special confluence of international competition, presidential directive, and necessary funding is very unlikely to be repeated in our lifetime."

Hubbard grants that other world nations see the moon as the key destination for human exploration. Not only the ESA but also Russia and China have announced that a human presence on the moon is part of their strategic space goals. "Clearly, countries that have never had boots on the moon wish to do so," says Hubbard. One possibility for the United States, he suggests, is low-cost lunar exploration, possibly by private enterprise, while at the same time continuing our quest to reach Mars.

Cost is always the issue. A recent study by NASA's Jet Propulsion Laboratory outlined a reasonably priced program for a human mission to Mars. The plan could only be accomplished, states the report, by ending NASA's contribution to the International Space Station as early as 2024 and no later than 2028. Next, human missions to Mars would proceed with a crewed landing on the Mars moon Phobos in 2033, followed by a short-stay landing on Mars in 2039, and a one-year stay in 2043. Each

"In the same way that we remember President Kennedy's challenge that motivated us to dream of reaching the moon, the leader of the nation who makes a commitment to land on Mars within two decades will be remembered throughout history."

—Buzz Aldrin

mission would build on previous ones, building a legacy, infrastructure, and capabilities for those that follow. That Mars architecture as scripted, independently priced by the nonprofit group The Aerospace Corporation, could be accomplished within the current NASA budget, adjusted for inflation. And one would assume—although it's not included in the internal study—that international partners and private enterprise will bear some of the cost.

Edge of deep space

WHILE THE QUESTION OF WHAT to do with the moon remains unresolved, the idea of getting a toehold on the space around Earth's moon is getting traction. It starts with Orion, a multipurpose spacecraft designed to carry human explorers on long-duration deep space missions. The first step appears to be to use an orbit around the moon as a proving ground for the technologies necessary to advance human exploration into the solar system.

Leading U.S. aerospace firms are part of the plan, being called on to develop life support systems, radiation protection, and communication technologies for cis-lunar, or moon-orbiting, missions—advancements that will likely help plot out the way to other deep space destinations: an asteroid, perhaps, and Mars soon after. Lockheed Martin Space Systems of Denver, Colorado—the builder of the Orion—is working to augment Orion's capabilities so that it can sustain a crew within a specialized habitat in moon orbit for 30 days or longer, and then allow them to leave that habitat uncrewed for a period of time until the next mission, explains Josh Hopkins, Lockheed Martin's space exploration architect.

Hopkins says that putting in place a cis-lunar habitat allows a phased approach in moving outward to Mars. "We're looking at getting something up and operating in cis-lunar space relatively soon," says Hopkins, but also working toward delivering new technology after that first launch: more advanced recycling systems and life support gear, for example, that will get a trial run in moon orbit prior to going to Mars. "Being on the edge of deep space at the moon is an important step," Hopkins says. "The moon is almost exactly a thousand times farther away than the International Space Station . . . and a Mars mission is about a thousand times farther away than the moon." One step at a time.

Beyond our backyard

IN 2006 14 SPACE AGENCIES from around the world established the International Space Exploration Coordination Group (ISECG), a deliberative body whose stated goal is to advance space exploration "through coordination of their mutual efforts."

The organization has circulated a "global exploration road map" meant to coordinate human and robotic space exploration of solar system destinations where humans may one day live and work.

"Space exploration enriches and strengthens humanity's future," states the opening passage of the key ISECG document. "Searching for answers to fundamental questions such as: 'Where did we come from?' 'What is our place in the universe?' and 'What is our destiny?' can bring nations together in a common cause." The road map describes a phased expansion of human reach beyond low Earth orbit with missions to Mars as the common long-term goal. "The human migration into space is still in its infancy," states the document. "For the most part, we have remained just a few kilometers above the Earth's surface—not much more than camping out in the backyard. It is time to take the next step."

The ISECG road map advocates missions into cis-lunar space and to the surface of the moon prior to venturing beyond, to Mars—a strategy that raises the moon-versus-Mars controversy. Some agencies around the world feel they can demonstrate critical capabilities needed for Mars by way of the moon, notes NASA's Kathy Laurini, co-chair of the road map's working group. "It's no secret that other space agencies want to put humans on the lunar surface on the path to Mars," she says. "NASA has said that we don't see the moon as a necessary step on the path to Mars for demonstrating the capabilities we would contribute."

Laurini adds that she is very aware there are those space agencies wanting to go to the moon. "We respect that and have told them that we would contribute to the missions on the path to Mars, but they need to do more than talk about them. They need to invest," she observes. There are significant scientific advances to be made by placing humans on the moon, such as developing ways of using resources found there. Lunar exploration could advance technologies needed for Mars. Surface power systems, surface mobility systems, surface habitations, a human ascent module: Technologies such as these on the moon could play forward to any Mars exploration. Laurini is adamant: International collaborations are essential to get us to Mars—and those benefits circle back to Earth as well. As the ISECG document concludes, "This new era of space exploration will strengthen international partnerships through the sharing of challenging and peaceful goals."

That sentiment was expressed eloquently by Susan Eisenhower, a distinguished international policy analyst with a special interest in the relations between the United States and Russia. Eisenhower was asked in April 2014 to testify before the U.S. Senate Subcommittee on Science and Space at its "From Here to Mars" hearing. Reflecting on the history of a contentious space race and the prospect of a new era of international collaboration, Eisenhower said, "As we know from history, it is always easier to terminate space cooperation than it is to get it started again. And

For 520 days, an international crew of astronauts lived a simulated Mars experience in a structure designed to include both a habitation and a region similar to the planet's surface. Here a member of the Mars500 project treads on reddish sand meant to resemble the surface of Gusev Crater, landing site of NASA's Spirit rover.

we will not be able to meet our long-term goals in space without it." Implicitly alluding to the "blue marble" effect—the new view of Earth from space, a unifying spectacle—she continued: "Space has unique capacities to serve the global community. It can be a force for preventive diplomacy, transparency and for sustaining and building bonds among those who are willing to put aside solely national pursuits."

Private sector missions

While nations thrash out their plans to make Mars a common goal, public-private collaborations are also in the works. NASA has contracted with Bigelow Aerospace to develop human spaceflight missions that leverage its innovative space habitat, called the B330. An expandable space module, it provides 12,000 cubic feet of space in a pressurized cabin and can support a crew of up to six people. Bigelow hopes to see B330 habitats used to support human spaceflight missions to the moon, Mars, and beyond.

Perhaps the most publicized Mars-bound enterprise so far is Mars One, a one-way flight plan considered by most in the industry to be an ultra-extreme long shot. Based in the Netherlands, Mars One is a nonprofit founded by Dutch visionaries Bas Lansdorp and Arno Wielders and dedicated to establishing a human settlement on Mars. Using social media to invite applications from individuals willing to take a one-way ticket to Mars, they received over 200,000 applications from all over the world. "That means the most popular job vacancy of all times was actually to go to Mars to stay," says Lansdorp. Out of those, they selected 100 people who will form international crews of four people each and move to Mars starting in 2026.

The Mars One challenge has never minced words: This is not a round-trip ticket. "A one-way mission to Mars greatly reduces the infrastructure needed. The absence of a return mission means that there is no need for a return vehicle, return propellant, or systems to produce the propellant locally, all of which would require a significantly larger amount of resources and technology development." Communications systems, rovers, and living units will be remotely delivered to Mars before the initial crew of four arrives. "The settlement will develop as those inhabiting it become architects of their own environment," the Mars One Web site explains. It remains to be seen whether this daring venture, developed without government affiliation and with few corporate partners, will get off the ground and onto the surface of Mars.

Musk on Mars

ELON MUSK MAY JUST MAKE IT. Entrepreneur and chief rocketeer at Space Exploration Technologies, or SpaceX, Musk is devoted to a distinct personal and company trajectory. The SpaceX Web site boldly tells the story: "SpaceX designs, manufactures and launches advanced rockets and spacecraft. The company was founded in 2002 to revolutionize space technology, with the ultimate goal of enabling people to live on other planets."

On Musk's time line, the SpaceX capsule called Red Dragon will travel unpiloted and land cargo on Mars in 2018 and in 2020, ahead of a humans-to-Mars mission departing Earth as early as 2024 for Red Planet arrival in 2025. Musk's plans for starting a city on Mars will necessitate a Mars colonial transporter.

Musk has said that his own personal spark of ambition came about in college as he was trying to sort out what realms of work could have a significant, positive effect on the future of humanity. "And the three things that I came up with," he told the National Press Club in 2011, "were the Internet; sustainable energy, both production and consumption; and then space exploration . . . specifically, making life multiplanetary. And I didn't expect when I was in college to actually be involved in all three of those areas." But indeed he has: from PayPal (Internet banking) to Tesla

DISASTERS

Power struggles | As more missions arrive, who is in charge? An emergency could cause chaos. Tensions over colony primacy, immigration issues, even differences of race or religion could tear through the fabric of a growing human settlement on Mars.

What could go wrong?

(electric cars) with a side track to solar energy and now SpaceX, the company with which he hopes to make life multiplanetary. In Musk's realm of cosmic consciousness, it's a requirement, he says "to design vehicles that can transport life over hundreds of millions of miles of irradiated space to an environment that they did not evolve to exist in."

Whether by choice or necessity, Musk thinks humans will ultimately become a multiplanetary species. We may someday need another planet to move to, he advises. And in the meantime, some people might choose to. "If you can reduce the cost of a flight to Mars, or moving to Mars, to around the cost of a middle class home in California"—around half a million dollars, in other words—"then I think enough people would buy a ticket and move to Mars to be part of creating a new planet and [founding] a new civilization," Musk told the Press Club audience. Pointing to Earth's population number of seven billion people now, rising to probably eight billion by the midpoint of the century, Musk said that even if one in a million people decided to do that, it adds up to 8,000 people. "And I think probably more than one in a million people would decide to do that," he added.

Filling out paperwork

BACKING MUSK'S VIEW, there appears to be no shortage of aspirant Marsnauts, relying on a U.S. government indicator. In February 2016, NASA announced that a record number of people responded to the space agency's call for astronauts. "It's not at all surprising to me that so many Americans from diverse backgrounds want to personally contribute to blazing the trail on our journey to Mars," said NASA administrator Charles Bolden.

More than 18,300 people filled out the paperwork to join NASA's 2017 astronaut class. That's nearly three times the number of applications received in 2012 for the most recent class, and it's a number that far surpasses the previous record of 8,000 applicants in 1978. This huge pool of applicant will be whittled down to a handful. A NASA astronaut selection board process ends with picking just 8 to 14 individuals who move into full-fledged astronaut candidate status. NASA expects to announce its choices in mid-2017.

The Red Planet calls. There are passionate people eager to sign on the dotted line, ready and willing to dedicate their lives to adventuring across time and space to another world. They are pilgrims riding atop a wave of technological progress. While Bolden hasn't promised the 2017 class of astronauts a ticket to Mars, he has said, "This next group of American space explorers will inspire the Mars generation to reach for new heights, and help us realize the goal of putting boot prints on the Red Planet." It's an invitation into an unknown and fascinating future. ■

UP AND AWAY

Lifting off from Kazakhstan in March 2016, the joint Europe-Russian ExoMars 2016 begins its seven-month journey to Mars. A month after takeoff, it sent back its first image, test-run for its complex imaging system.

esa
БРИЗ-М
ПРОТОН-М

INDIA MAKES IT TO MARS

Scientists and engineers congratulate one another (below) at the Indian Space Research Organization in Bangalore as Mangalyaan—"Mars craft" in Hindi—reached Mars orbit on September 24, 2014. "History has been created today," said Prime Minister Narendra Modi, and many agreed from launch date on (right) as their country became the first to reach Mars on its first try.

Dinesh Gup
Advocate High Cou
Time : 7.00 to 9.00 p.m.
लोकमत
मंगळ झेप!
मंगळाची वारी...
'रिक्षा'च्या खर्चात
सादर आहे नवा
ENO इनो रिफ्रेशिंग जेल व टॅब्स
ENO
1981
अवघे कोलकाता झाले 'सचिन'मय
मनाई आहे

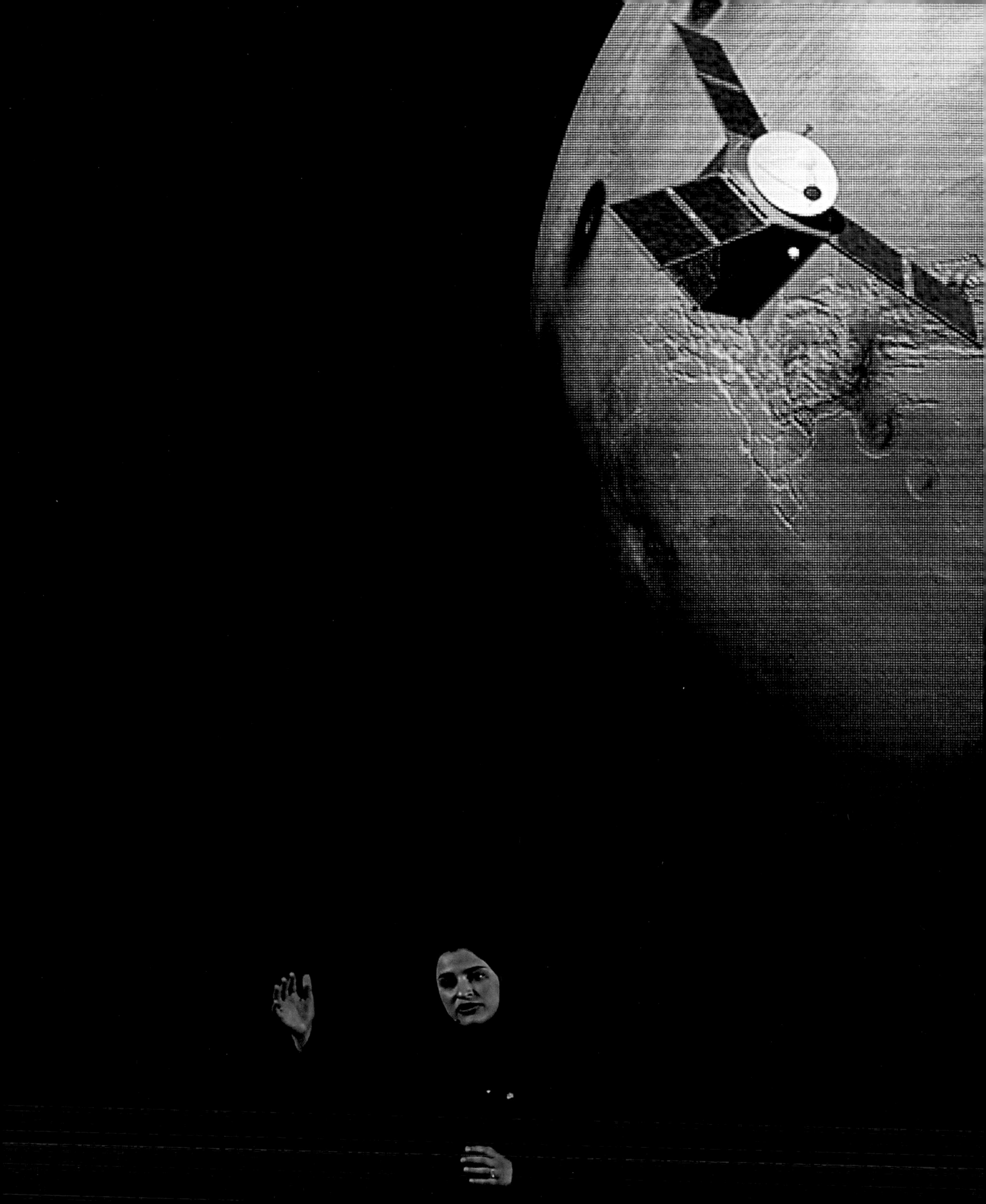

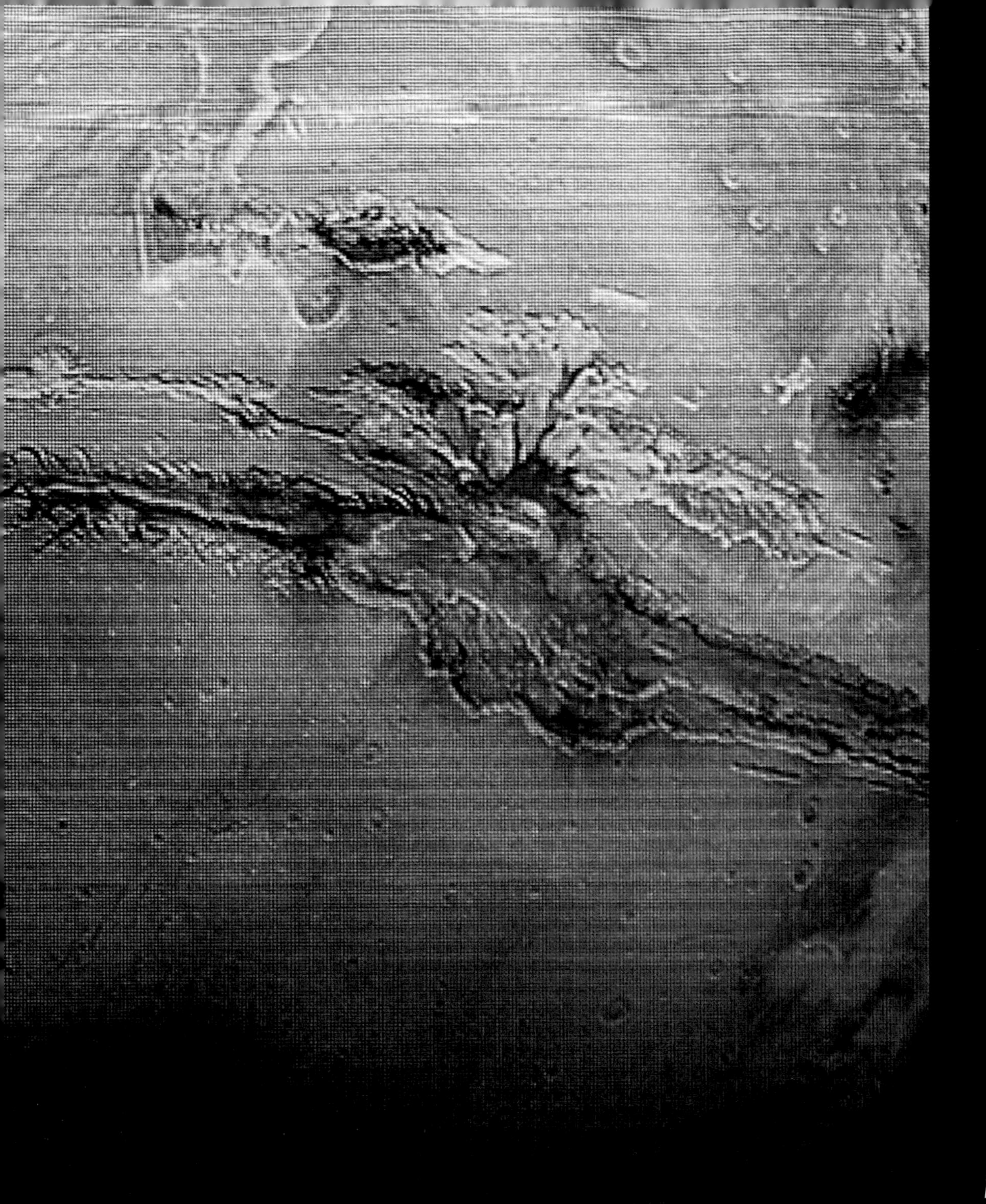

ANNOUNCING A UAE MARS MISSION

Sarah Amiri, deputy project manager of a United Arab Emirates Mars mission called al-Amal ("hope" in Arabic), describes plans for the unmanned probe arriving at Mars in 2021 and orbiting at least two years, collecting data on the planet's atmosphere.

DRACONIAN MANEUVERS

From 2012 on, SpaceX craft have been making unmanned round-trip deliveries of supplies to the International Space Station, launching (as in 2014, right) from the Cape Canaveral Air Force Station in Florida. In January 2016 the company revealed video of a successful hover test (below), a key step toward landing a vessel carrying humans without parachuting it into open ocean.

FRONT ROW SEATS

The next challenge to be met by the visionary engineers of SpaceX is to send humans into low Earth orbit and beyond—all the way, if CEO Elon Musk has his way, to Mars. Here is a glimpse of the interior of *Crew Dragon,* a SpaceX craft designed to carry up to seven astronauts.

S-IV
UNITED

Space Policy Decision-Maker

HEROES | JOHN LOGSDON

Professor Emeritus, Space Policy Institute, The George Washington University

Just how global is the political appetite to reach for Mars? According to space policy sage John Logsdon, "This must and will be an international coalition." Even so, realistically, Logsdon adds, only the United States can lead the universal league of the willing. "The United States still spends more money on civil space activities than the rest of the world combined. So it makes no sense for some other country to have the ambition to lead this effort. Only the United States has the resources."

Professor emeritus of political science and international affairs at the George Washington University in Washington, D.C., Logsdon has been an astute and highly valued voice in space policy decision-making over the decades, recently contributing to a new plan for affordable human missions to Mars.

Still, there's anxiety in certain quarters that a humans-on-Mars project will never get the commitment necessary to support a long-term and expensive effort. Logsdon disagrees. "The United States did a sustained program for 40 years called the space shuttle, from 1972 through 2011. Also, the United States sustained the International Space Station from 1982 out to now 2024. Since Apollo, the government has sustained, between the space shuttle and the space station, a relatively steady level of effort." To the professor, these are "existence proofs" that the American government will provide the resources to maintain an expensive space effort "as long as the program's pace and ambition match the likely funding."

Author of several books on space exploration, including seminal looks at President John F. Kennedy's decision to go to the moon, Logsdon says: "There seems to be a pretty broad consensus that humans-to-Mars is the appropriate goal for the U.S. space program." That said, "Those who advocate a Mars program have to be ready when the timing seems propitious," sort of like running in place. He agrees, as NASA's chief, Charles Bolden, has said, that NASA is closer to boots on Mars than ever before. "That doesn't mean that we're close," Logsdon adds. "In my mind, the path to Mars leads through the moon." He believes that an international effort to revisit the moon has to precede any international mission to Mars.

In 2010 President Barack Obama declared that the United States was going to Mars. "That remains the guiding policy," grants Logsdon, but adds that succeeding U.S. presidents will need to keep that goal in view in order to make it happen. "It's almost bipolar," Logsdon says. "Either the U.S. continues on a path toward deep space, cis-lunar, and then onward to Mars . . . or it's the end of the government human spaceflight program. There's really nothing else to do."

Our quest for the stars and planets dates back to the days of the Cold War space race between the United States and the Soviet Union. Here Wernher von Braun, leading rocket engineer of those days, explains the Saturn booster system to President John F. Kennedy, whose dedication to space exploration spurred Americans to the moon.

INFLATION FACTOR

Headquartered in North Las Vegas, Nevada, Bigelow Aerospace specializes in the design and manufacture of inflatable habitats including BEAM, the Bigelow Expandable Activity Module. Inflatables can be part of an orbiting assembly or serve as habitats on faraway worlds such as Mars.

ONE LAUNCH AT A TIME

The long haul to Mars starts with stepping-stone success. Here, United Launch Alliance, a joint venture of Boeing and Lockheed Martin, readies the powerful Delta IV Heavy booster in December 2014 to hurl an unpiloted Orion craft spaceward. Builder of the Orion, Lockheed Martin calls the successful launch "our first step on a journey to Mars."

WELCOME TO MY ROCKET

British entrepreneur Sir Richard Branson proudly introduces the Virgin Galactic SpaceShipTwo in February 2016. The dawn of private space travel, beginning with pay-per-view trips on his company's suborbital rocket plane, is near at hand.

LAUNCH, LAND, REPEAT

Space-age billionaire Jeff Bezos, of Amazon.com fame and fortune, has been making steady progress with his company, Blue Origin, and is gaining recognition as a contender in the competition to pioneer privately funded space travel. In January 2016 its New Shepard rocket repeated a second successful takeoff and safe vertical landing (right)—a noteworthy accomplishment deserving of celebration (below).

BLUE ORIGIN

Canadä

HEROES | MARCIA SMITH

Editor, SpacePolicyOnline.com; President, Space and Technology Policy Group

If a humans-to-Mars program is to live long and prosper, many stars have to align—not in celestial terms but in terms of politics and policy. "NASA has been through the trenches enough times to know that a presidential pronouncement counts only if there is sufficient money to implement it," says Marcia Smith, founder and editor of SpacePolicyOnline.com and president of Space and Technology Policy Group in Arlington, Virginia. "That is the enduring obstacle. Sending people to Mars with a reasonable amount of risk is very expensive."

Smith is a space policy aficionado—former director of the Space Studies Board and of the Aeronautics and Space Engineering Board at the U.S. National Research Council. "Sending people to Mars, as a government project, has significant support in Congress today," she points out, "as evidenced by the budget increases NASA has gotten over the past two years. But even with those increases, the amount of money that will be needed year after year after year seems almost insurmountable."

As a long-term strategy, NASA has adopted the Evolvable Mars Campaign for achieving human missions to the planet that can adapt to changing circumstances. "It is a realistic approach that is challenged by those in the space program who yearn for a return to the Apollo era," Smith comments. "They want a specific architecture leading to people on the Mars surface in the early-mid-2030s, insisting that it will be impossible to win support without a date certain and specific plans."

Could the emergence of private space endeavors, Elon Musk's SpaceX in particular, help make a human trek doable? Yes, says Smith, the private sector—meaning the commercial sector acting in a commercial manner, not as a government contractor—could play a critical role. Right now, SpaceX gets loads of money from the government. It has avoided the label of "government contractor" because of its nontraditional contracts, but the government is the source of the money nonetheless, she explains.

"How far any of these entrepreneurial companies is willing to go without government money to develop their systems . . . is an open question," Smith adds. Without a doubt, any humans-to-Mars plan that involves reasonable risk will take a long time, a lot of money, and a lot of talent. "That means a collaborative effort among multiple governments and the private sector," says Smith.

The question remains as to whether the goal is to be the first, or to have a long-term program of dozens or hundreds or thousands of people going there indefinitely, Smith observes. "That is a much bigger challenge with different motivations. How many want to be second? Or tenth? Or the one-hundredth person? To me, those are the real explorers and [that] involves a long-term, international/commercial step-by-step approach."

Symbolizing today's public-private partnerships, the SpaceX Dragon resupply craft pulls into close proximity to the International Space Station as crew inside the station grapple it with a robotic arm.

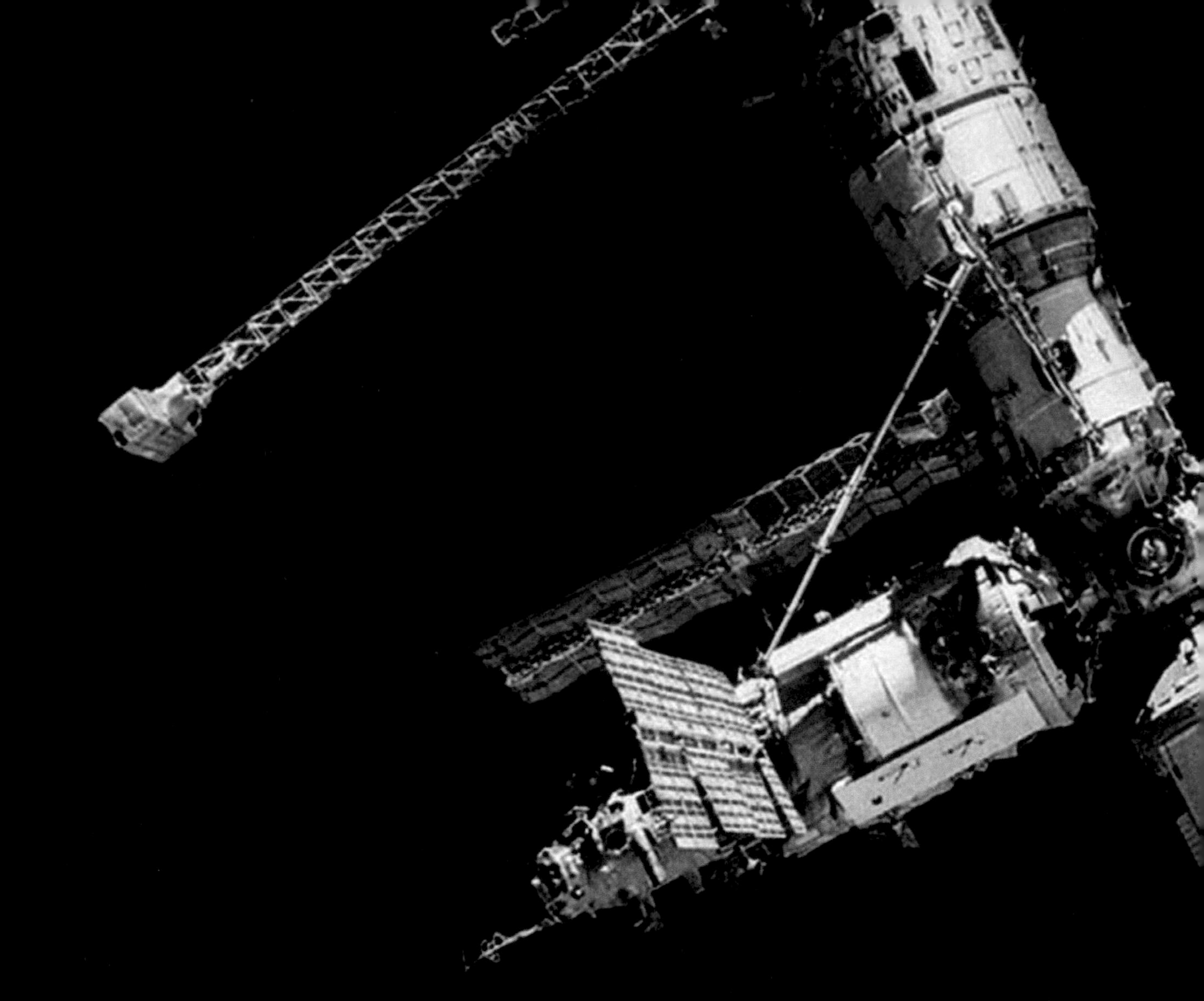

COUPLING OF CULTURES

With the Cold War space race a thing of the past, NASA and the Russian space agency entered into a new era of international cooperation in June 1995 as the space shuttle *Atlantis* docked with the Russian space station Mir—first of 11 space shuttle visits.

NEW INSIGHT ON MARS

NASA's InSight (Interior Exploration using Seismic Investigations, Geodesy and Heat Transport, a mission to plumb Mars's geological depths) is now set for launch in 2018, delayed from an original 2016 date because of problems with a prime instrument in the lander's science payload. When it finally lands on the planet, the spacecraft will deploy as shown here.

WALLS OF VICTORIA CRATER

In 2007 NASA's Mars exploration rover Opportunity sent back this revealing image of Cape St. Vincent, a promontory jutting out from the walls of Victoria Crater. Opportunity and its companion rover, Spirit, landed on opposite sides of Mars in January 2004. Many years later, Opportunity is still exploring the Red Planet.

The children born on Mars will know no other landscape, no other way of life. They will face very different challenges, and yet human nature will endure.

MARSL

AND

As we look into the future on Mars, our imaginations take off in many directions. This habitation design, called Ice House, includes inflatable windows infused with radiation-shielding gas attached to walls made with native ice.

MARS
MULTIPLE TOURS AVAILABLE
ROBOTIC PIONEERS / ARTS & CULTURE / ARCHITECTURE & AGRICULTURE

6

Marsland

NVISIONING A HALF CENTURY of Mars settlement is fortune-telling at its best. Surely it's too simpleminded just to picture gated communities of air-locked and domed enclaves. Danger is always present on the boundaries of exploration, but that hasn't stopped us before and it won't stop us this time. In the words of T. S. Eliot, "Only those who will risk going too far can possibly find out how far they can go." Microgravity, long-duration travel, radiation, cosmic rays: Will these fear factors be surmounted? And will they be replaced by others we cannot foretell from today's vantage point? We need to prepare ourselves psychologically for a huge transformation, say many of the experts.

Think about it, says B. J. Bluth, emeritus faculty member in sociology at California State University, Northridge: When people moved from Europe to the New World and then on to the Far West, the people and their cultures changed. "Attitudes, values, and ways of living underwent significant alteration . . . The same phenomenon will affect those who opt to settle the planets." And it's not just psychological, she continues: Space pioneers will develop physical, immunological, cultural, and social differences from their forebears who stay on Earth.

Starting from the assumption that we will not only land on Mars but establish communities there, we need to reckon with the biological effects of long-term life and, ultimately, birth on the fourth planet. First and foremost are the effects of radiation. A primary health concern for humans on Mars is radiation exposure. Astronauts on a Mars trek will be exposed to galactic cosmic radiation and increased radiation during periodic solar storms. "The biggest threat from radiation exposure is the possibility of dying from radiation-induced cancer sometime after a safe return to Earth," says Ron Turner, a space radiation specialist at ANSER, a research institute in Falls Church, Virginia. Limited research also suggests that radiation

Will the day come when posters promote tours, events, and exhibits on Mars? Working with strategists at NASA's Jet Propulsion Laboratory, graphic designers illustrate the principle "If you create it, it will happen."

EPISODE 6

CROSSROADS

In the wake of the damage wrought by the storm, the situation begins to look bleak for the Mars mission. Nervous investors and national governments have become convinced that the dangers—financial, physical, and psychological—have become too great to justify the extension of the human presence on the planet. Despite the best efforts of dedicated advocates back on Earth, the curtain looks set to fall on humanity's presence on Mars . . . until an unexpected discovery reveals that the biggest surprises on Mars have yet to be found.

exposure could lead to effects that manifest during a long-duration mission instead of years afterward, adds Turner. Degenerative or acute effects could include heart disease, reduced effectiveness of the immune system, and even, potentially, Alzheimer's-like neurological symptoms.

"The space radiation environment will be a critical consideration for everything in the astronauts' daily lives, both on the journeys between Earth and Mars and on the surface," explains Ruthan Lewis, an architect and engineer with the human spaceflight program at NASA's Goddard Space Flight Center in Greenbelt, Maryland. According to Sheila Thibeault, a materials researcher at NASA's Langley Research Center in Hampton, Virginia, one promising cutting-edge shielding idea is use of hydrogenated boron nitride nanotubes. Known as hydrogenated BNNTs, researchers have successfully made yarn out this nanotube material. It's flexible enough to be woven into the fabric of space suits, she reports, providing astronauts strolling on Mars with radiation protection even out on the unkind Martian surface.

The gravitational force that humans will experience on Mars is roughly three-eighths (0.375) that on Earth. Very little is known about what the long-term effects of a lessened tug of gravity will be on human health. Life science experiments done on the International Space Station do show the loss of bone mass. That weakening may be a dilemma for crews reintroduced to a gravity environment after crossing the lengthy expanse of space to Mars. "The magnitude of this [effect] has led NASA to consider bone loss an inherent risk of extended space flights," says Jay Shapiro, team leader for bone studies at the National Space Biomedical Research Institute in Houston, Texas.

The gravity on Mars is not your best friend, suggests Kevin Fong, associate director at the Centre for Altitude, Space, and Extreme Environment Medicine at University College London. An anesthesiologist and physiologist, Fong is the author of *Extreme Medicine: How Exploration Transformed Medicine in the Twentieth Century.* Mars's lower gravity will present a cascade of medical issues to be addressed: Worries about bone density, muscle strength, and the body's circulation pattern must be taken into account, he explains in an article for *Wired* magazine. "Deprived of gravitational load, bones fall prey to a kind of spaceflight-induced osteoporosis. And because 99 percent of our body's calcium is stored in the skeleton, as it wastes away, that calcium finds its way into the bloodstream, causing yet more problems from constipation to renal stones to psychotic depression," he explains.

Lower gravity means a body grows taller as well. Space travelers who have returned to Earth are up to two inches taller. That's because there is less gravity pushing down on the vertebrae so they stretch out several inches. The condition doesn't last long once they are back on Earth, and they return to their Earth-gravity height. All in all, though, changes in bone strength, muscles, and the immune system are only a few of the alterations to the human body that will occur for people inhabiting Mars.

And once generations inhabit the Red Planet, what does the future hold for someone born on Mars? It is possible that a Mars-born human could find travel back to the crushing gravity of Earth unbearable. "We have no data on kids at anything but one gravity," says Al Globus, a senior research engineer at San Jose State University and a member of the National Space Society's board of directors. "So, what will happen at one-third gravity on Mars? We don't know."

But, Globus says, "there is one thing we do know with a great deal of confidence: Kids on Mars will grow up weak compared to those growing up on Earth. Bone and muscle develop in response to stress. On Mars, the gravitational stress is much less, so bone and muscle will be weaker."

Reclamation project

WHILE MARS HAS THE POWER to change our bodies, what can we do to morph Mars and make it Earth-like? The concept of terraforming—reconstructing the climate and surface of Mars to make it comfortably habitable for humans—has been under discussion for years now. This is a bigger, longer-term, and more fantastical project than installing a few habitation modules in which exploratory crews can live for a few months or a year: Terraforming means turning the entire planet of Mars into one that can support Earth's kind of life.

In the past there have been some slam-bang schemes like smacking the planet with water-laden comets, or orbiting it with huge mirrors to reflect sunlight to raise the surface temperature, or sprinkling the polar caps with dark surface material siphoned off the Martian moons, another way to turn up the thermostat. Then there's the proposal to spread genetically altered microbial life in the form of dark-colored lichens, algae, or bacteria, a biological way to soak up sunlight and warm the Martian atmosphere.

To engineer Mars into a livable planet, we will need to provide three essentials: accessible water, breathable oxygen, and a livable climate. NASA space scientist Christopher McKay has given those tasks a methodical look and pulled together a terraforming time line. The first Mars warming phase would likely take a hundred years.

"The primary challenge to making Mars a world suitable for life is warming that planet and creating a thick atmosphere. A thick, warm atmosphere would allow liquid water to be present and life could begin," McKay notes. While he adds that warming an entire planet may seem like a concept from the pages of science fiction, in fact we are demonstrating this capability on Earth now. "By increasing the carbon dioxide content of the Earth's atmosphere and the addition of super-greenhouse gases, we are causing a warming on Earth that is on the order of many degrees centigrade per century. Precisely these same effects could be used to warm Mars," McKay points out.

On Mars we could purposefully produce super-greenhouse gases and rely on carbon dioxide released from the Martian polar caps and absorbed in the ground. The result would be a thick, warm atmosphere blanketing the Red Planet. McKay adds that warming of several degrees per century here on Earth is happening now without a focused effort. Thus, the timescale for deliberately warming Mars through super-greenhouse gas production is shorter, he reasons.

Photosynthetic organisms can be introduced as conditions warm, and organic biomass will begin to flourish. As a natural consequence, the biomass will begin consuming the nitrates and perchlorates in the Martian soil, ultimately generating nitrogen and oxygen. As this 100-year stretch of terraforming takes hold, temperature and pressure increases will bring forth liquid water at equatorial and midlatitudes on Mars. Frequent snow and occasional rainfall will occur as rivers flow and equatorial lakes form. Eventually a hydrological cycle on the planet will be established similar to that found in the Dry Valleys of Antarctica. Tropical trees can be planted; insects and some small animals can be introduced. Humans will still need gas masks to provide oxygen and prevent high levels of carbon dioxide in the lungs.

In the next phase, to allow humans to breathe naturally, McKay foresees an oxygenation period taking much longer, moving toward oxygen levels above 13 percent and carbon dioxide levels below one percent at sea-level atmospheric pressure. On Earth,

The human imagination has been on Mars for decades. As early as 1953, in this and many other illustrations, Chesley Bonestell, American pioneer of space art, envisioned a human presence and the technology involved on the Red Planet.

the efficiency of the global biosphere in using sunlight to produce biomass and oxygen is 0.01 percent. The timescale for producing an oxygen-rich atmosphere on Mars, McKay says, thanks to plants spread over the surface of the planet that exhibit that efficiency, is 10,000 times 17 years, or roughly 100,000 years. There may be methods and technologies as yet unknown that we may be able to use to speed the process, he adds: "In the future, synthetic biology and other biotechnologies may be able to improve on this efficiency." The time line nevertheless still stretches far into the future.

While 100,000 years is a bit of a wait, there is no harm in sharpening our terraforming skills, McKay argues, by conducting tiny experiments with big consequences for Mars. A plant germination trial test on a robotic lander on Mars, for example, could help us design ways to use photosynthesis to produce oxygen for ourselves. But plans to garden on Mars raise a much bigger question, McKay advises: the impact of Martian life-forms on our own efforts to terraform the planet.

If there is no life on Mars, then the situation is relatively straightforward. But proving the absence of something is something of a conundrum, and even after extensive exploration, it may be hard to conclude that life is completely absent on Mars rather than simply not present at the specific locations investigated. But if any form of life is discovered, McKay continues, then we will need to carefully define the relationship between it and the Earth life-forms we want the planet to support. It may be that life-forms found on Mars will be related to those on Earth, possibly

"We are all . . . children of this universe. Not just Earth, or Mars, or this System, but the whole grand fireworks. And if we are interested in Mars at all, it is only because we wonder over our past and worry terribly about our possible future." —Ray Bradbury

due to meteorite exchange millennia ago. But if forms of life are discovered that are unrelated to Earth life, then not only technical but also huge ethical issues emerge, he concludes.

A walk in the park

TERRAFORMING THE ENTIRE PLANET—a full Mars makeover—is a multigenerational project under discussion and debate among the experts today. In the meantime, others are proposing that we select certain areas of the planet's surface to work on, amounting to a network of seven parks designed to preserve different regions on the Red Planet. This notion has been championed by Charles Cockell, a professor of astrobiology at the University of Edinburgh in Scotland.

Mars is home to stretches of desert, splendid canyons, extinct shield volcanoes, and expansive polar ice caps. By preserving portions of these features, a diversity of planetary parks with different features of outstanding beauty and intrinsic natural worth could be established, says Cockell. The parks would also allow for maximum preservation of scientific heritage, geologically, and perhaps biologically. Regions with special human significance could also be preserved: the first human landing site, spots where robotic vehicles achieved special milestones, and even the remains, still operating or dead, of crafts that made it to Mars ahead of humans.

Gerda Horneck, at the Institute of Aerospace Medicine at the German Aerospace Center in Cologne, Germany, sees the initiative as analogous to national park systems right here on Earth. Presenting their ideas in the prestigious journal *Space Policy,* she and Cockell see Martian parks as a necessary part of the planet's human future, a response to what they call "the inevitable development of industry and tourism on Mars." Park regulations could regulate nearby industrial development and "might become the focus of tourist visits, just as the preservation of the Grand Canyon National Park on Earth, for instance, is made possible by encouraging people to visit and appreciate its splendour and special status."

If parks, why not museums? Even today, humanity has already left traces there. "It is one of the few planets besides our own which contains our artifacts," observes Beth O'Leary, professor emeritus of anthropology at New Mexico State University in Las Cruces. "There have been both successful and unsuccessful missions to Mars, although more than half of all launched missions ended in failure," she points out, encouraging us to "think about the preservation of historic spacecraft for future visitors, ourselves or perhaps others on Mars."

Successful robotic landers on Mars represent the evolution of planetary travel and science and have the unusual characteristic of being self-documenting, O'Leary notes. "They returned imaging and other information about the Martian surface to

Earth. How the design, planning, and cultural aspects of spacecraft launched toward Mars changed over the last half century is provided, in part, by the material record. These artifacts are historically significant in the exploration of space."

These legacy leftovers also document what Dirk Spennemann, a futurist and associate professor of cultural heritage in New South Wales, Australia, has called "robotic culture"—a heritage part human and part machine. "Our lunar heritage is rife with all kinds of robotics beginning in the earliest stages to the present. The sites on Mars, so far, are in a cultural landscape modified by robots." Failed missions and their crash sites are of equal importance in any forensic investigation of Mars, says O'Leary. Identifying the sites where the events occurred and where physical evidence could still remain is critical. Clues exist, she continues, to figure out the reasons and nature of the malfunction, boosting the opportunity for successful Mars landings in the future. Some of this valuable research can continue in person once humans are on the scene.

Copernican perspective

For decades now, since astronauts have gained the vantage point of outer space and sent back photographs taken of Earth, millions have marveled at the "big blue marble" and been humbled by the "pale blue dot" imagery taken by the robotic Voyager 1 from a distance of about 3.7 billion miles, all evoking a symbol and sense of planetary unity. Called the "overview effect" in a book by the same name by author Frank White, the concept has inspired a nonprofit foundation dedicated to keeping that symbol alive and its meaning vital.

That view of Earth from outer space has an impact unlike any possible from the ground: "The experience of seeing firsthand the reality of the Earth in space, which is immediately understood to be a tiny, fragile ball of life, hanging in the void, shielded and nourished by a paper-thin atmosphere," as the Overview Institute Web site describes it. "From space, the astronauts tell us, national boundaries vanish, the conflicts that divide us become less important, and the need to create a planetary society with the united will to protect this 'pale blue dot' becomes both obvious and imperative. Even more so, many of them tell us that from the Overview perspective, all of this seems imminently achievable, if only more people could have the experience!" We had a taste of it in early 2014, when Curiosity sent back images of Earth from Mars: the brightest object in the sky yet still a tiny dot.

For millennia, White says, we had no overview effect—no concrete way to experience the reality that we live on a planet hurtling through the universe at a high rate of speed. Once astronauts went into orbit and to the moon, they brought it into stark reality for all of us. By contrast, points out White, humans arriving at Mars will be

Mars500, the long-term simulation of a living situation on Mars created by the European and Russian space agencies, could only begin to evoke the experience. Here a visitor peers in on the simulated world.

already versed in the overview effect. If you are on the moon, you can imagine yourself making a trip back to Earth if you had to. But if you are on Mars, a return trip will be less likely. It could take as long as it did for the early colonists to go back to England, or the pioneers to return East—and it's likely to be expensive as well. It might require a much more difficult transition to adjust physically from Mars back to Earth than it has to go from the moon back to Earth. "Depending on what the expectations are, this could be very difficult for some of the settlers to accept," adds Frank White.

Then Earth will become like the Red Planet is to those here on Earth today: a tiny light in the sky, devoid of any distinguishing features without the help of a telescope. The message from many comes through loud and clear: The pioneer travelers to Mars will not be coming back. Their very journey begins with an attitude of self-sufficiency and independence. "The Martians will soon develop their own culture and seem like true 'aliens' to Earthlings," envisions White, leading ultimately to a "declaration of independence" from Earth by Mars.

White also cites his discussions with Oregon-based philosopher Nick Nielsen. For millennia humans had only what Nielsen calls the "homeworld effect," and we did not experience the reality that we live on a planet hurtling through the universe at a high rate of speed. When the astronauts went into orbit and to the moon, they had a direct experience of our planet's real situation.

"By contrast," says White, "when people go to Mars, they will start with an overview of that planet. They will know intuitively what 'Spaceship Mars' is, though it has taken a long time to understand 'Spaceship Earth.' With all the spacecraft currently circling Mars, we already have an overview, though it is from afar.

"I believe that this will make our early 'Martians' a lot more comfortable with Mars than with Earth, which will seem so remote to them," White advises. "We also need to think about a permanent reduction in gravity and its impact on the thought processes of the people on Mars, and especially on their children. I think they will evolve quickly—mentally, emotionally, and biologically."

The thrill of it all

Is the peopling of Mars a hyped-up, hypervelocity selling job? Rayna Elizabeth Slobodian, a master of education candidate at York University's Department of Anthropology in Toronto, Ontario, Canada, recently published what she called an "anthropological critique of the rush to colonize Mars."

"We have powerful and influential people in the media continuing the conversation regarding Mars colonization," Slobodian says, be it Elon Musk at SpaceX, Apollo 11 moon walker Buzz Aldrin, or superstar astronauts Chris Hadfield and Scott Kelly. While these space advocates may know what they are talking about, the Mars craze that has swept the general public, symbolized in the social media chatter surrounding Mars One, might be an indicator of less attractive social characteristics, says Slobodian. "Marketers selling colonization to the public include perspectives such as biological drives, species survival, inclusiveness and utopian ideals," she writes, arguing that "much of our desire to colonize space within the next decade is motivated by ego, money and romanticism," including a quest for immortality. "We shouldn't gloss over the risks to people's lives in order to sell ideas of utopia or inspiration," she warns.

Nevertheless, the excitement is palpable. People do think going to Mars will be fun—maybe more fun, in fact, if done here on Earth through virtual reality headgear or as a vacation destination. To answer those holiday dreams, plans are afoot to build Mars World in Las Vegas, Nevada: an immersive Red Planet experience promised to be as spectacular as any other theme park on Earth. Experience one-quarter Earth gravity, putz around in a rover, stroll the space walkway, or lounge in the Mars-themed spa—all contained in a dome "as big as the Giza Pyramid" and "so large, the

DISASTERS

Homesick too long | As a Martian settlement develops independent from the home planet, social isolation evolves into a deep sense of hopeless disconnectedness. There is never a chance to step outside for a breath of fresh air.

What could go wrong?

Rose Bowl could fit inside." Chief designer John Spencer, founder of the Space Tourism Society, grants that the venue will build on the "longstanding connection between science fiction and entertainment and real things." In short, he says, it's not science fiction but "science future." Leave it to Las Vegas to offer a trip to Mars without ever leaving Earth.

We are the Martians

VERY FEW PEOPLE REALIZE that the first monument to be taken by humans to Mars already exists, along with instructions on where this bit of stainless steel is to be placed. An 8-by-10-inch plaque, it is temporarily on display at the Smithsonian's National Air and Space Museum in Washington, D.C., next to a full-size engineering model of the Viking Mars lander. The inscription on the plaque reads, "Dedicated to the memory of Tim Mutch, whose imagination, verve, and resolve contributed greatly to the exploration of the Solar System."

Thomas A. "Tim" Mutch was a leading space scientist, planetary geologist, and leader of the Viking imaging team responsible for the cameras that operated on the first U.S. spacecraft to land on Mars, in 1976, and send pictures back home to Earth. Based on those images, Mutch published his third book, *The Martian Landscape*. Mutch was also an avid mountaineer who organized numerous high-altitude expeditions. He was quoted as telling his students and fellow climbers, "I can't take you all to Mars, but I can show you a little of what it means to explore."

Mutch died on one of those expeditions, on the slopes of Mount Nun in the Himalaya in 1980. A year after his death, NASA renamed the Viking 1 lander, then still operating on Mars, the Thomas A. Mutch Memorial Station. At that time NASA administrator Robert Frosch unveiled the stainless steel plaque as well, announcing that it was to be affixed to the side of the Viking 1 lander by the first team of explorers to reach Chryse Planitia, the site on Mars where the robotic craft still sits.

By the time that happens, many other rovers, landers, and other Earth paraphernalia will have touched down on the surface of Mars. But nothing will be as thrilling, as historic, as death defying, as life and future affirming as the day when human beings step onto the planet, calling Mars their own. Those future homesteaders of Mars will likely remember the words spoken by Ray Bradbury, author of *The Martian Chronicles* and longtime visionary of space travel and exploration.

In 1976, Bradbury rejoiced in the touchdowns of the Viking 1 and 2 robotic landers on Mars, saying: "Today we have touched Mars. There is life on Mars, and it is us . . . extensions of our eyes in all directions, extensions of our mind, extensions of our heart and soul have touched Mars today. That's the message to look for there: 'We are on Mars. We are the Martians!' " ■

ORBITING IMAGINATION

The further into the future our minds take us, the more we depend on artists' visions rather than photography to represent our future on Mars. Here an illustration envisions a deep space exploration vehicle dubbed Kronos 1 as it orbits above Valles Marineris.

METROPOLIS DOWN UNDER

Some plans propose that the planetary surface itself should be used to protect a human habitation from the harsh climate and high radiation levels found on Mars. If that turns out to be the plan, we might end up creating an entire bustling underground, with city lights glowing day and night.

HOUSING OF THE FUTURE

Multiple competitions have inspired visionary habitations combining science, art, and engineering: the rigors of a Martian environment, the graceful lines of architectural design, and the technology to unite the two. Mars Ice House (below), first-place winner in NASA's recent 3-D–printed habitat competition, makes building materials out of Mars water, fiber, and aerogels. LavaHive (right), third-place winner, proposes a modular design that uses a novel casting technique to meld Martian soil and space vehicle parts together.

tech
SAS
SENTRY AIR SYSTEMS, INC.

HEROES | EUGENE BOLAND

Chief Scientist, Techshot

Not many companies can claim a "Mars room" designed for experiments requiring an ecosystem paralleling that on Mars. Techshot's test chamber allows scientists to do just that: mimic the Martian atmospheric pressure, shifts in day-night temperatures, and the ruthless solar radiation that beats down on the Red Planet's surface. In the midst of this fabricated ecosystem, chief scientist Eugene Boland is researching "ecopoiesis"—the concept of initiating life in a new place and, more precisely, the creation of an ecosystem capable of supporting life. It's not terraforming—modifying the soil, air, and atmosphere to make it more habitable for humans—but it is a creative step forward in sorting out methods to prolong human stays on Mars.

Boland and fellow researchers are appraising the feasibility of using ecosystem-building pioneer organisms to churn out oxygen by using Martian soil. Some organisms within the test-bed experiment also could remove nitrogen from the Martian soil. "This is a way to produce oxygen," Boland says, explaining that it would cut out the need to haul weighty gas canisters to Mars at great cost. "Our microbes will be utilizing the entire Martian environment, the soil plus subterranean ice, plus atmosphere to make breathable oxygen. Let's send microbes and let them do the heavy-lifting for us." In due course, large biodomes on Mars that enclose ecopoiesis-provided oxygen through bacterial or algae-driven conversion systems could be cozy quarters for expeditionary teams, Boland says.

Boland's research has been backed by the NASA Innovative Advanced Concepts Program. The proposal is to have special devices carried aboard a future Mars rover. At selected sites, the small container-like gear would be augured into the ground a few inches deep. Then selected Earth organisms in the container—extremophiles like certain cyanobacteria—will interact with the Mars soil that has been brought into the apparatus. Once in operation, the device will sense the presence or absence of a metabolic product like oxygen and report the finding to a Mars-orbiting relay satellite.

The container's design includes a tight seal, necessary to prevent the Earthly organisms from being exposed to the Martian atmosphere. Boland sees the experiment as the first major leap from laboratory studies to an experimental (as opposed to analytical) on-the-spot Mars inquiry of great interest to planetary biology, ecopoiesis, and terraforming.

Boland's vision: Biological oxygen factories on Mars that are kick-started with microbes—a method, he believes, that can yield the supplemental oxygen that will be needed. "I'm a biologist and an engineer," says Boland. "So I want to put those two things together to make a useful tool" to solve a known problem: how to provide the oxygen essential for human life on Mars.

Ensuring adequate oxygen is a key step in sending humans to Mars. Scientists on Earth are now evaluating oxygen-producing techniques that could be transported and established on the Red Planet.

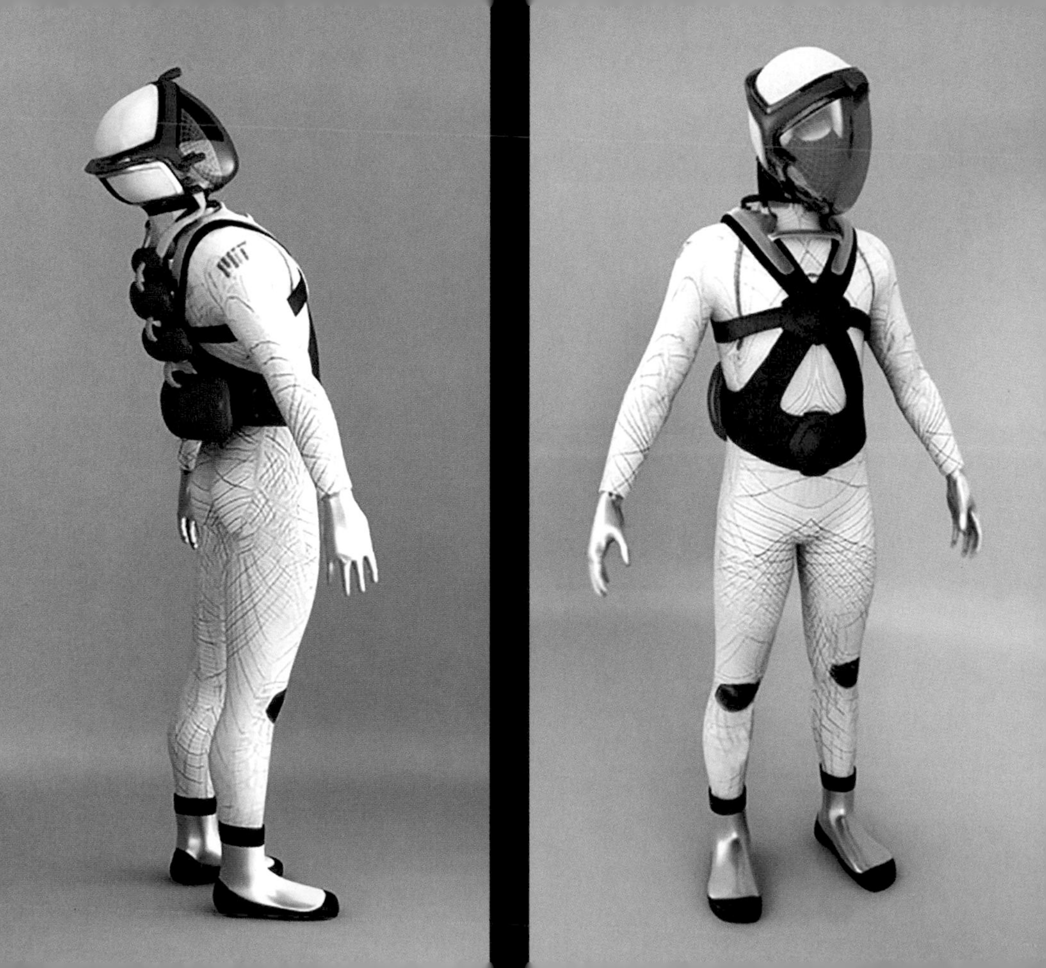
MIT

ORGANIC COLUMNS

To create this underground habitation, mining robots will precede humans at the site and probe the planet's surface to find basalt, an igneous rock known to be present on Mars. The strongest columns of basalt found become foundation pillars. Interior details are made of basalt roving, a woven filament already used in the aerospace industry.

began, but somehow more than a million people squeeze life from this parched land. Such a realization makes the prospect of terraforming a barren planet like Mars (right) more feasible.

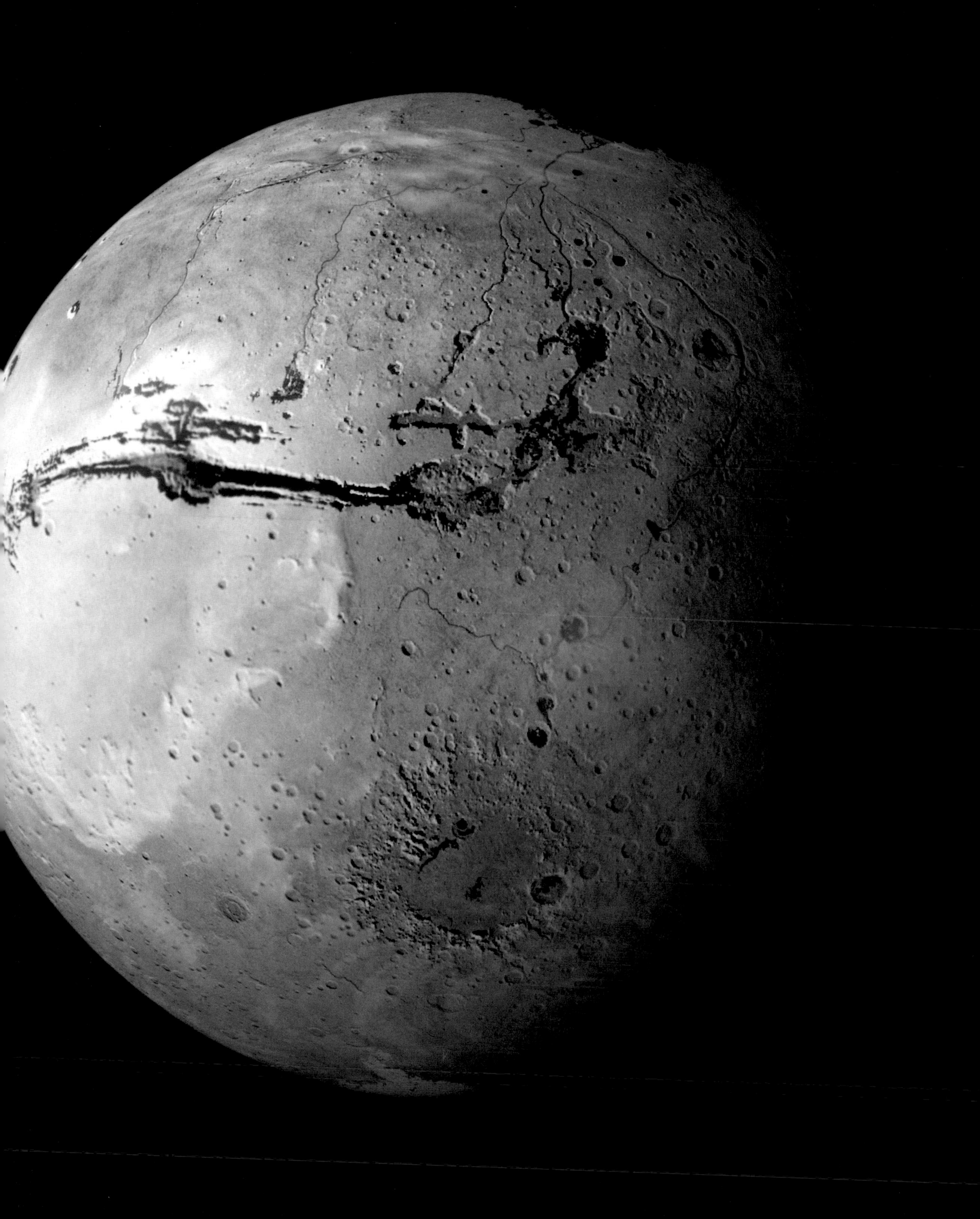

ALL EYES ON MARS

The world will be watching as the first humans land on Mars—just as they were for the first steps onto the moon in 1969 and for the successful landing of NASA's Curiosity rover on Mars in August 2012. From Times Square (shown here) to Tokyo, millions of people cheered Curiosity on. Estimates put the number watching the landing online as 3.2 million.

E. ALDRIN

HEROES | BUZZ ALDRIN

Former Astronaut and Space Exploration Advocate

If you're trying to mobilize a movement to Mars, it's best to ask advice from someone who is already a deep space pioneer—someone like Buzz Aldrin. Aldrin was a member of the Apollo 11 team, the first manned landing crew on the moon in July 1969. He was second to set foot on Earth's natural satellite, after fellow astronaut Neil Armstrong, during the epic voyage that made human history.

Now, more than 45 years later, Aldrin is forceful in his conviction that Mars is humanity's next destination. He sees the Red Planet as "a waiting island in the blackness of space." "Personally, I have lived through a lot of firsts," Aldrin says. Firsts in science and exploration, he continues, "require a special kind of leadership, the sort that is defined by courage." Placing people on the surface of Mars is a continuation of humankind's unending legacy of exploration.

Buzz Aldrin has been busy of late designing a system he calls Cycling Pathways to Occupy Mars. "Cycling Pathways is an engineering approach, technically sound and ready to be put into practice, not a destination. The physics is all there," he explains. "Moreover, continuing refinements at Purdue, MIT, and at my Buzz Aldrin Space Institute at Florida Tech confirm that if we start now, successful human landings for continuous occupancy can reach Mars by 2040."

As Aldrin says—and technical papers confirm—existing cycler architecture emphasizes reusability, with resupply for low-cost transfers of crews. "This really can be done," he points out. "The earthbound equivalent would be the economies created by ferryboats ferrying passengers back and forth across rivers."

Aldrin's plan calls for cycling spacecraft and two or more Mars landers capable of relaunching to intercept a cycler swinging by from Earth for a nine-month transfer to Mars or its moon Phobos. The Mars base incorporates improved technologies, he continues, to be developed earlier at a similarly designed base on the moon. A telerobotically constructed Mars base, controlled from Phobos, would draw on lessons learned through the assembly of an international lunar base from a U.S. spacecraft stationed in lunar orbit, Aldrin suggests.

"Choosing to lead with various cycler systems not only puts America back to the forefront of human space exploration," he says, "but offers a significant way to bring together all other spacefaring nations . . . to a worldwide sharing for the greatest human endeavor in history.

"The U.S. president who appeals to our higher angels and takes us closer to the heavenly body we call Mars will not only make history. He or she will long be remembered as a pioneer for humanity to reach, to comprehend, and to settle Mars." Challenges Aldrin: "And if not now, when? And if not us, who? This is our time . . . this is your time!"

In a photograph that made history, Apollo 11 astronaut Buzz Aldrin stands on a stark landscape during the first human landing on the moon in July 1969. Reflected in his visor are the photographer—his fellow astronaut, Neil Armstrong—and their lunar module, the *Eagle*.

SHIFTING SANDS

NASA's Curiosity rover transmitted this picture of the Bagnold Dunes along the northwestern flank of Mount Sharp. Observations over time have shown that these sand dunes can move up to a yard a year. Images from Curiosity are really streams of data to be analyzed by NASA scientists; this photo is color-adjusted to make the dunes appear as they might in Earth daylight.

MARTIAN SUNSET

NASA's Mars Exploration Rover Spirit captured this twilight view from Gusev Crater on May 19, 2005, the rover's 489th sol, or Martian day, on the planet.

SCENES FROM A SILENT PLANET

As envisioned by artist Julien Mauve, humanity's footsteps are leading to Mars—a totally different world from ours, an enigmatic globe full of wild landscapes, mystery, dangers, and promise.

TIME LINE | INTERNATIONAL MARS MISSIONS

1960

USSR
Marsnik 1 (Mars 1960A)
10 October 1960
Attempted Mars Flyby
(Launch Failure)
Marsnik 2 (Mars 1960B)
14 October 1960
Attempted Mars Flyby
(Launch Failure)

1962

USSR
Sputnik 22
24 October 1962
Attempted Mars Flyby
Mars 1
1 November 1962
Mars Flyby (Contact Lost)
Sputnik 24
4 November 1962
Attempted Mars Lander

1964

USA
Mariner 3
5 November 1964
Attempted Mars Flyby
Mariner 4
28 November 1964
Mars Flyby
USSR
Zond 2
30 November 1964
Mars Flyby (Contact Lost)

1965

USSR
Zond 3
18 July 1965
Lunar Flyby, Mars Test Vehicle

1969

USA
Mariner 6
25 February 1969
Mars Flyby
Mariner 7
27 March 1969
Mars Flyby
USSR
Mars 1969A
27 March 1969
Attempted Mars Orbiter
(Launch Failure)
Mars 1969B
2 April 1969
Attempted Mars Orbiter
(Launch Failure)

1971

USA
Mariner 8
9 May 1971
Attempted Mars Flyby
(Launch Failure)
USSR
Cosmos 419
10 May 1971
Attempted Mars Orbiter/Lander
Mars 2
19 May 1971
Mars Orbiter/ Attempted Lander
Mars 3
28 May 1971
Mars Orbiter/ Lander
USA
Mariner 9
30 May 1971
Mars Orbiter

1973

USSR
Mars 4
21 July 1973
Mars Flyby (Attempted Mars
Orbiter)
Mars 5
25 July 1973
Mars Orbiter
Mars 6
5 August 1973
Mars Lander (Contact Lost)
Mars 7
9 August 1973
Mars Flyby (Attempted Mars
Lander)

1975

USA
Viking 1
20 August 1975
Mars Orbiter and Lander
Viking 2
9 September 1975
Mars Orbiter and Lander

1988

USSR
Phobos 1
7 July 1988
Attempted Mars Orbiter/
Phobos Landers
Phobos 2
12 July 1988
Mars Orbiter/Attempted Phobos
Landers

1992

USA
Mars Observer
25 September 1992
Attempted Mars Orbiter (Contact
Lost)

1996

USA
Mars Global Surveyor
7 November 1996
Mars Orbiter
RUSSIA
Mars 96
16 November 1996
Attempted Mars Orbiter/Landers
USA
Mars Pathfinder
4 December 1996
Mars Lander and Rover

1998

JAPAN
Nozomi (Planet-B)
3 July 1998
Mars Orbiter
USA
Mars Climate Orbiter
11 December 1998
Attempted Mars Orbiter

1999

USA
Mars Polar Lander
3 January 1999
Attempted Mars Lander
Deep Space 2 (DS2)
3 January 1999
Attempted Mars Penetrators

2001

USA
2001 Mars Odyssey
7 April 2001
Mars Orbiter

2003

EU
Mars Express
2 June 2003
Mars Orbiter and Lander
USA
Spirit (MER-A)
10 June 2003
Mars Rover
Opportunity (MER-B)
8 July 2003
Mars Rover

2005

USA
Mars Reconnaisance Orbiter
12 August 2005
Mars Orbiter

2007

USA
Phoenix
4 August 2007
Mars Scout Lander

2011

RUSSIA
Phobos-Grunt
8 November 2011
Attempted Phobos Lander
CHINA
Yinghuo-1
8 November 2011
Attempted Mars Orbiter
USA
Mars Science Laboratory
26 November 2011
Mars Rover

2013

INDIA
Mangalyaan
5 November 2013
Mars Orbiter
USA
MAVEN
18 November 2013
Mars Scout Mission Orbiter

2016

EU
ExoMars 2016
14 March 2016
Mars Orbiter and Lander

2018 on

USA
NASA Mars Rover, 2020
Next Mars Orbiter, 2022
China
Mars Orbiter/Lander/Rover
2020
ESA
ExoMars Rover, 2020
UAE
Hope Mars Orbiter, 2020
Japan
Mars Moon Exploration
2022

Source: Chronology of Mars Exploration, NASA Space Science Data Coordinated Archive

Acknowledgments

I want to extend my thanks to the countless individuals and organizations I contacted during the writing of this book—too numerous to list. Without your assistance, this book would not have been possible.

My appreciation goes out to Mars planners—NASA's Rick Davis, Jr., and Steve Hoffman at SAIC—who provided invaluable insight and comment to help shape this volume.

Many thanks to my wife, Barbara, who made sure I remained grounded on Earth while simultaneously being on the Red Planet.

A special salute to my early Mars Underground colleagues, especially Chris McKay, Carol Stoker, Carter Emmart, Ben Clark, Penny Boston, Steve Welch, Buzz Aldrin, Keli McMillen, and the late Tom Meyer, a guiding light and convener of all things possible.

I am also indebted to the book's Mars Team, specifically the editing (and deadline-taunting) talents of Susan Tyler Hitchcock, the artful eye of photo editor Katherine Carroll, and the masterly design of the book by David Whitmore.

Last, gratitude is extended to the global cadre of people who are tirelessly building the bridge between Earth and Mars.

—Leonard David

About the author

Leonard David has been reporting on space exploration for more than five decades. He is a winner of the 2010 National Space Club Press Award and a past editor in chief of the National Space Society's *Ad Astra* and *Space World* magazines. Together with Buzz Aldrin he co-authored *Mission to Mars: My Vision for Space Exploration.* David writes the "Space Insider" column at Space.com and serves as a contributing writer for several other publications. He lives in Golden, Colorado, with his wife, Barbara.

ILLUSTRATIONS | CREDITS

Cover, National Geographic Channels/Brian Everett; Back Cover, NASA/United Launch Alliance; Front Flap, NASA/JPL; Back Flap (UP) Barbara David; Back Flap (LO), Jeff Lipsky; 1, Reproduced courtesy of Bonestell LLC; 3, Lockheed Martin; 4, NASA/JPL-Caltech/University of Arizona; 10-11, NASA/JPL-Caltech/University of Arizona; 18-9, NASA/Goddard Space Flight Center Scientific Visualization Studio; 19, NASA/JPL-Caltech; 20-21, NASA/JPL-Caltech; 22, NASA/JPL-Calech/University of Arizona; 24, National Geographic Channels/Robert Viglasky; 29, NASA; 31, NASA/JPL-Caltech/University Arizona/Texas A&M University; 32-3, Lockheed Martin/United Launch Alliance; 34-5, NASA/United Launch Alliance; 36-7, NASA/Bill Ingalls; 38, NASA/Aerojet Rocketdyne; 39, NASA/Stennis Space Center; 40-41, NASA/JPL-Caltech/University of Arizona; 42, NASA/JPL-Caltech; 42-3, NASA; 44-5, NASA/JPL/University of Arizona; 46-7, NASA/JPL-Caltech; 48-9, NASA/JPL-Caltech; 49, NASA/JPL-Caltech; 50-51, NASA/JPL-Caltech/MSSS; 52, NASA/JPL-Caltech/Malin Space Science Systems; 53, NASA/JPL; 54-5, NASA/JPL-Caltech; 56-7, NASA/JPL-Caltech; 60-61, NASA/JPL/Arizona State University; 62-3, NASA/Goddard Space Flight Center Scientific Visualization Studio; 62 (LE), NASA/JPL-Caltech; 62 (RT), NASA; 64-5, NASA; 66, NASA/Bill White; 68, National Geographic Channels/Robert Viglasky; 76-7, British Antarctic Survey; 78, French Polar Institute IPEV/Yann Reinert; 78-9, ESA/IPEV/PNRA–B. Healy; 80-81, ESA/IPEV/ENEAA/A. Kumar & E. Bondoux; 82-3, Neil Scheibelhut/HI-SEAS, University of Hawaii; 84-5, Oleg Abramov/HI-SEAS, University of Hawaii; 85, Christiane Heinicke; 86, NASA; 87, Carolynn Kanas; 88-9, NASA; 90-91, EPA/NASA/CSA/Chris Hadfield; 91, NASA; 92-3, NASA; 94-5, Phillip Toledano; 96, NASA/Bill Ingalls; 97 (UP), NASA/Robert Markowitz; 97 (LO), NASA/Robert Markowitz; 98, IBMP RAS; 98-9, ESA—S. Corvaja; 100-101, Mars Society MRDS; 102-103, Mars Society MRDS; 103, Mars Society MRDS; 104, NASA; 105, Jim Pass; 106-107, ESA/DLR/FU Berlin–G. Neukum, image processing by F. Jansen (ESA); 107, ESA/DLR/FU Berlin; 108-109, NASA/JPL/ASU; 110-11, NASA/Goddard Space Flight Center Scientific Visualization Studio; 111 (UP LE), NASA/JPL-Caltech; 111 (UP RT), NASA; 111 (LO), NASA/JPL/University of Arizona; 112-13, NASA/JPL/University of Arizona; 114, NASA/JPL-Caltech/MSSS; 116, National Geographic Channels/Robert Viglasky; 119, NASA/Wallops BPO; 123, NASA/Emmett Given; 125, NASA/Artwork by Pat Rawlings (SAIC); 126, Percival Lowell (PD-1923); 127, NASA; 128-9, NASA Langley Research Center (Greg Hajos & Jeff Antol)/Advanced Concepts Lab (Josh Sams & Bob Evangelista); 130-31, Kenn Brown/Mondolithic Studios; 132, NASA/Bill Stafford/Johnson Space Center; 133, NASA/Bill Stafford and Robert Markowitz; 134-5, © Foster + Partners; 136, NASA/Artwork by Pat Rawlings (SAIC); 137, Jim Watson/AFP/Getty Images; 138-9, NASA/Bigelow Aerospace; 140, Data: MOLA Science Team; Art: Kees Veenenbos; 140-141, NASA/JPL/USGS; 142-3, NASA/Ken Ulbrich; 144, Haughton-Mars Project; 145, SETI Institute; 146-7, Bryan Versteeg/Spacehabs.com; 148-9, NASA/JPL/University of Arizona; 149, NASA/JPL-Caltech/Univ. of Arizona; 150-151, NASA/Goddard Space Flight Center Scientific Visualization Studio; 151 (UP LE), NASA/JPL-Caltech; 151 (UP RT), NASA; 151 (CTR), NASA/JPL/University of Arizona; 151 (LO), Carsten Peter/National Geographic Creative; 152-3, Carsten Peter/National Geographic Creative; 154, Joydeep, Wikimedia Commons at https://en.wikipedia.org/wiki/Cyanobacteria#/media/File:Blue-green_algae_cultured_in_specific_media.jpg (photo), http://creativecommons.org/licenses/by-sa/3.0/legalcode (license); 156, National Geographic Channels/Robert Viglasky; 159, NASA/JPL-Caltech/Univ. of Arizona; 163, NASA/JPL-Caltech/Cornell/MSSS; 166-7, Carsten Peter/National Geographic Creative; 168-9, Carsten Peter/National Geographic Creative; 170, Mark Thiessen/National Geographic Creative; 171, Image Courtesy of the New Mexico Institute of Mining and Technology; 172-3, Trista Vick-Majors and Pamela Santibáñez, Priscu Research Group, Montana State University, Bozeman; 174-5, ESA; 176, DLR (German Aerospace Center); 177, George Steinmetz/National Geographic Creative; 178-9, Kevin Chodzinski/National Geographic Your Shot; 179, Diane Nelson/Visuals Unlimited; 180-81, NASA/JPL/Ted Stryk; 182, Wieger Wamelink, Wageningen University & Research; 182-3, Jim Urquhart/Reuters; 184, NASA/JPL-Caltech/Lockheed Martin; 185, Paul E. Alers/NASA; 186-7, DLR (German Aerospace Center); 188-9, ESA; 190, NASA/JPL-Caltech/Cornell/MSSS; 190-191, NASA/JPL-Caltech/MSSS; 192, NASA; 193,

NASA; 194-5, NASA/JPL-Caltech/Univ. of Arizona; 196-7, NASA/Goddard Space Flight Center Scientific Visualization Studio; 197 (UP LE), NASA/JPL-Caltech; 197 (UP RT), NASA; 197 (CTR), NASA/JPL/University of Arizona; 197 (LO LE), Carsten Peter/National Geographic Creative; 197 (LO RT), ESA/J. Mai; 198-9, ESA/J. Mai; 200, Official White House Photo by Chuck Kennedy; 202, National Geographic Channels/Robert Viglasky; 205, Andrew Bodrov/Getty Images; 209, ESA/IBMP; 212-13, ESA–Stephane Corvaja, 2016; 214, Reuters/Abhishek N. Chinnappa; 214-15, Punit Paranjpe/AFP/Getty Images; 216-17, AP Photo/Kamran Jebreili; 218, Trey Henderson; 218-19, SpaceX; 220-21, SpaceX; 222, NASA; 223, David M. Scavone; 224-5, NASA/Bill Ingalls; 226-7, Lockheed Martin; 228-9, Al Seib/Los Angeles Times/Getty Images; 230, Blue Origin; 230-31, Blue Origin; 232, NASA; 233, Courtesy Marcia Smith; 234-5, NASA; 236-7, NASA/JPL-Caltech/Lockheed Martin; 238-9, NASA/JPL/Cornell; 240, SEArch/CloudsAO; 240-41, NASA/Goddard Space Flight Center Scientific Visualization Studio; 241 (UP LE), NASA/JPL-Caltech; 241 (UP RT), NASA; 241 (CTR), NASA/JPL/University of Arizona; 241 (LO LE), Carsten Peter/National Geographic Creative; 241 (LO RT), ESA/J. Mai; 242-3, SEArch/CloudsAO; 244, Courtesy NASA/JPL-Caltech; 246, National Geographic Channels/Robert Viglasky; 249, Reproduced courtesy of Bonestell LLC; 253, Natalia Kolesnikova/AFP/Getty Images; 256-7, Maciej Rebisz; 258-9, Alexander Koshelkov; 260, Team Space Exploration Architecture/Clouds Architecture/NASA; 260-61, LavaHive Consortium; 262, Techshot, Inc.; 263, Photo from Eugene Boland courtesy of Practical Patient Care magazine; 264, Dr. Dava Newman, MIT: BioSuit™ inventor; Guillermo Trotti, A.I.A., Trotti and Associates, Inc. (Cambridge, MA): BioSuit™ design; Michal Kracik: BioSuit™ helmet design; Dainese (Vincenca, Italy): Fabrication; Douglas Sonders: Photography; 265 (LE), Dr. Dava Newman, BioSuit™ inventor/Guillermo Trotti, Trotti Studio, BioSuit™ design/Michal Kracik, BioSuit™ helmet design; 265 (RT), Dr. Dava Newman, BioSuit™ inventor/Guillermo Trotti, Trotti Studio, BioSuit™ design/Michal Kracik, BioSuit™ helmet design; 266-7, ZA Architects; 268, DEA/C. Dani/I. Jeske/Getty Images; 268-9, Data: MOLA Science Team; Art: Kees Veenenbos; 270-1, Navid Baraty; 272, NASA/Neil A. Armstrong; 273, Rebecca Hale/National Geographic Staff; 274-5, NASA/JPL-Caltech/MSSS; 276-7, NASA/JPL/Texas A&M/Cornell; 278-9, Julien Mauve.

MAP CREDITS

Mars Hemispheres Maps (pages 6-9, 12-15)
Base Map: NASA Mars Global Surveyor; National Geographic Society.
Place Names: Gazetteer of Planetary Nomenclature, Planetary Geomatics Group of the USGS (United States Geological Survey) Astrogeology Science Center *planetarynames.wr.usgs.gov.*
IAU (International Astronomical Union) *iau.org.*
NASA (National Aeronautics and Space Administration) *nasa.gov.*

East Melas Proposed Human Exploration Zone (EZ) Map (page 29)
Data from: "Landing Site and Exploration Zone in Eastern Melas Chasma," A. McEwen, M. Chojnacki, H. Miyamoto, R. Hemmi, C. Weitz, R. Williams, C. Quantin, J. Flahaut, J. Wray, S. Turner, J. Bridges, S. Grebby, C. Leung, S. Rafkin LPL, University of Arizona, Tucson, AZ 85711 (mcewen@lpl.arizona.edu), University of Tokyo, PSI, Université Lyon, Georgia Tech, University of Leicester, British Geological Survey, SwRI-Boulder.
THEMIS daytime-IR mosaic base map: NASA/JPL/Arizona State University/THEMIS.

Potential Exploration Zones Map (pages 58-59)
Data assembled by Dr. Lindsay Hays, Jet Propulsion Laboratory-Caltech.
Topography Base Map: NASA Mars Global Surveyor (MGS); Mars Orbital Laser Altimeter (MOLA).

INDEX

Boldface indicates illustrations.

A

Abbud-Madrid, Angel 29–30
Additive manufacturing (3-D printing)
food 121
habitats 122, 123, 124, **134–135, 260**
space suits 133
tools and equipment 121, **123**
Aerodynamic decelerators 25, 53
Aerojet Rocketdyne engines **38,** 39
The Aerospace Corporation 137, 207
Aging, premature 68
Aldrin, Buzz 16, 206, **272,** 273, **273**
Alzheimer's disease, and space radiation 68
America Makes 122
Amiri, Sarah **216–217**
Andromeda Strain (Crichton) 156
Ansari, Anousheh 125
ANSER (research institute) 245
ANSMET (Antarctic Search for Meteorites) 119
Antarctica
ANSMET expedition 119
extremophiles **172–173**
Lake Hoare **192**
as Mars analog 71–72, 75, **76–81, 152–153**
Mount Erebus **152–153, 168–169**
Apollo missions
Apollo 11 16, **272,** 273
Apollo 13 (film) 16
Apollo 17 156
compared to Mars mission 30–31
cost 205
risks from samples 156
Aquifers, on Mars 117, 159, 162, 164
ARADS (Atacama Rover Astrobiology Drilling Studies) 162–163
Archer, Doug 157–158
Armstrong, Neil 16, 273
Astronauts, NASA's call for (2016) 211
Astrosociology Research Institute 105
Atacama Desert, Chile 162–163, 268, **268**
Atlantis (space shuttle) **234–235**
Atlas V rocket **34–35**
Atmosphere, on Mars
carbon dioxide 115–116
change over time 202–203
oxygen 115–116, 248–249, **262,** 263
terraforming 248
water vapor 162
Aurora Australis **80–81**

B

B330 (space habitat) 209
Bacteria and viruses *see* Microbes
Bagnold Dunes, Mars **274–275**
Balloons 118, **119**
Basalt 266
BEAM (Bigelow Expandable Activity Module) **224–225**
Bezos, Jeff **230**
Bigelow Aerospace 209, **224–225**
Bigelow Expandable Activity Module (BEAM) **224–225**
Binsted, Kim 74–75
Biofilms 69
BioSuit **264–265,** 265
Blue Origin **230**
Bluth, B. J. 245
Boeing, United Launch Alliance **226–227**
Boland, Eugene 263, **263**
Bolden, Charles 26, 211, 223
Bone loss 246–247
Bonestell, Chesley, illustrations by **1, 249**
Boston, Penelope **170,** 171, **171**
Bowie, David 91
Bradbury, Ray 250, 255
Branson, Sir Richard **228–229**
British Antarctic Survey *see* Halley Research Station
Brunt Ice Shelf, Antarctica 77

C

Canals, on Mars **126–127**
Cancer, and space radiation 68, 115, 245
Cape Canaveral, Florida
Atlas V rocket **34–35**
Delta IV Heavy rocket **32–33**
SpaceX launches **200, 218–219**
Cape St. Vincent, Mars **238–239**
Carberry, Chris 203–204
Carbon dioxide, on Mars 115–116
Cargo Lander Zones 117
Cary, Craig **152–153**
The Case for Mars (Zubrin and Wagner) 72
Caves of Crystals, Mexico **166–167**
Center for Rapid Automated Fabrication Technologies 121–122
Center for Space Resources, Colorado School of Mines, Golden, Colorado 29
Center for the Advancement of Science in Space 125
Centre for Altitude, Space, and Extreme Environment Medicine 247
China, space program 201, 205, **205**
Christoforo, Samantha **91**
Chryse Planitia, Mars 255
Cis-lunar missions 207
Climate, terraforming Mars 248
Clouds AO (Clouds Architecture Office) 122
Cockell, Charles 251
College of William and Mary 159
Colonization of Mars *see* Habitations, on Mars; Human mission on Mars; Marsland
Colorado School of Mines, Center for Space Resources 29
Commercial sector efforts 209–211, **228–229,** 233
Communications delays 74, 105
Concordia (Italian-French base in Antarctica) 71–72, **78–79, 80–81**
Conley, Catharine "Cassie" 163–164, 185, **185**
"Contour crafting" 122
Cooties *see* Microbes
Coprates Chasma, Mars **194–195**
Cowley, Aidan 123
Crew Dragon (SpaceX craft) **220–221**
Crichton, Michael 156
Crystals, Cave of, Mexico **166–167**
Cueva de Villa Luz, Mexico **170**
Cuevas, Janine 39, **39**
Curiosity rover
Bagnold Dunes image **274–275**
capabilities 54
drilling samples **190–191,** 193
Earth image, from Mars 252
landing 25, **48–49,** 53, 270, **270–271**
landing site 8–9, 54
launch **34–35**
Mount Sharp image **114**
perchlorate detection 157
preparation **54–55**
selfies **50–51, 52**
"Sheepbed" rocks **163**
team 155
Cyanobacteria **154,** 155
Cycler architecture 273
Cycling Pathways to Occupy Mars 273

D

Davis, Rick 117
Decelerators, aerodynamic 25, 53
Deimos (Martian moon) 202
Delta IV Heavy rocket **32–33, 226–227**
Design Reference Architecture 5.0 137
Deutsches Zentrum für Luft- und Raumfahrt (German Aerospace Center) 69, 155, **186–187,** 251
Devon Island, Northwest Territories, Canada 72–73, **144,** 145
Disasters, potential
basic necessities 124
homesickness 254
hostile life-forms 164
power struggles 210
thrown off course 30
unknown terrain 74
Discovery (space shuttle), crew **104**
Dragon resupply craft **232**
Drake, Bret 137, **137**
Dunes, on Mars **4, 56–57,** 57
Dust devils **148–149**

E

Earth, view from space 209, 252–254
"Earth out of view" phenomenon 87
Eastern Hemisphere North, map 6–7
Eastern Hemisphere South, map 8–9
Ecopoiesis 263
Eisenhower, Susan 208–209
Eliot, T. S. 245
Emotional stresses *see* Mental and emotional stresses, on Mars
Endurance Crater, Mars 163, **163**
Entomopter (robotic bug) 118
Erebus, Mount, Antarctica **152–153, 168–169**
Erickson, James 125
ESA *see* European Space Agency
Europa (Jupiter's moon) **180–181**
European Space Agency (ESA)
British Antarctic Halley Research Station 71, **76–77**
Concordia (Italian-French base in Antarctica) 71–72, **78–79, 80–81**
European Service Module 204
ExoMars mission **188–189, 198–199,** 201–202, **212–213**
LavaHive 123–124, **260–261**
Mars500 mission 69, **98–99, 209, 253**
see also specific missions and vessels
Evolution of life 70
Evolvable Mars Campaign (NASA) 233
ExoMars mission **174–175, 188–189, 198–199,** 201–202, **212–213**
Exploration Flight Test-1 32, **32–33**
Explore Mars 203–204
Extreme Medicine (Fong) 247
Extremophiles
Antarctica 168, **172–173**
Atacama Desert, Chile 162–163
Cave of Crystals, Mexico **166–167**
Cueva de Villa Luz, Mexico **170**
producing oxygen on Mars 263

F

First Landing Site/Exploration Zone Workshop for Human Missions to the Surface of Mars 25
Fong, Kevin 247
Food
growing on ISS 165
growing on Mars 67, 119, 121, **146–147,** 165, **182–183**
Foster + Partners 119, 122–123
Frogs **178–179,** 179
Frosch, Robert 255

G

Gagarin Cosmonaut Training Center, Russia **36–37**
Gale Crater, Mars 54, 157

Ganges Chasma, Mars **10–11**
Garvin, James 155
Geographic features, of Mars 115
Geology, of Mars **184**
Georgia Institute of Technology 159, 161
German Aerospace Center (Deutsches Zentrum für Luft- und Raumfahrt) 69, 155, **186–187,** 251
Global vision 196–239
    China 201
    collaborations 201–204, 207–209, **209, 212–213**
    Europe 201–202
    India 202, **214–215**
    Japan 202
    private sector missions 209–211
    time line 280
    United Arab Emirates 202–203, **216–217**
    United States 202, 203, 204, 223
Globus, Al 247
Goddard Space Flight Center 155, 246
Grand Canyon of Mars *see* Valles Marineris, Mars
Grand Prismatic Spring, Yellowstone National Park, U.S. **177**
Gravity
    artificial 73
    on Mars 246–247
    microgravity 68
Grazer, Brian 16, 17
Green, James 25
Greenhouse gases 248
Greenhouses **146–147,** 165
Grimm, Robert 162
Gusev Crater, Mars 46, **209, 276–277**

H

Habitation Zone 117
Habitations, on Mars 110–125
    architecture 122–125, **136,** 137
    commuter architecture 118
    demands 110
    design competitions 122–125, **134–135,** 203, **260–261**
    Earth independence 121
    environmental difficulties 115
    growth of encampment 121, **130–131,** 137
    Ice House 122, **242–243, 260**
    inflatables 123–124, **134–135, 224–225, 242–243**
    infrastructure 121–122, **130–131**
    LavaHive 123–124, **260–261**
    native Martian design 122–125
    oxygen sources 115–116
    possible sites 124
    pressurized rovers **125**
    resources 121–122
    site possibilities 117
    space walks 118–119
    underground **258–259, 266–267**
    water sources 116–117
    zones of action 117
    see also Marsland
Hadfield, Chris **90–91,** 91
Hale Crater, Mars **159**
Halley Research Station, Antarctica 71, **76–77**
Hartmann, William 158–159
Haughton Mars Project (HMP) 72–73, 145
Hawaii Space Exploration Analog and Simulation (HI-SEAS) 74–75, **82–83, 84–85**
Haystack Observatory, MIT 115–116
Head, Jim 30–31
Healey, Beth **78–79**
Health impact of long-term space travel 67–69
Health risks on Mars
    bone loss 246–247
    perchlorates 156–157
    radiation 115, 245–246
    water safety 163–164
    see also Mental and emotional stresses, on Mars
Heat shields **142–143,** 143
Hecht, Michael 115–116
Heinicke, Christiane 82
HI-SEAS (Hawaii Space Exploration Analog and Simulation) 74–75, **82–83, 84–85**
High-resolution imaging science experiment (HiRISE) camera system 116–117, **149**
HMP (Haughton-Mars Project) 72–73, 145
Hoare, Lake, Antarctica **192**
Hoffman, Stephen 117, 119
Home base *see* Habitations, on Mars
Homesickness 254
Homeworld effect 253
Hopkins, Josh 207
"Hoppers," robotic 118
Horneck, Gerda 251
Housing *see* Habitations, on Mars
Howard, Ron, foreword by 16–17
Hubbard, Scott 205
Human mission on Mars
    benefits 105
    crew autonomy 74–75, 105
    crew selection 72–73
    first outpost, location for 24
    gravity and Mars-born humans 247
    JPL program for 205, 207
    landing technology **42–43**
    map of potential exploration zones 58–59
    Mars One 209–210
    nimble equipment, need for 73
    phased approach 207, 208
    pioneering mission 23, 53
    private sector efforts 209–211
    simulation research 69
    tourism poster **245**
    see also Habitations, on Mars; Mental and emotional stresses, on Mars
Hydrated minerals 30, 159, 161
Hydrogenated boron nitride nanotubes (BNNTs) 246
Hypersonic inflatable technology **42–43,** 53

I

Ice
    extremophiles **172–173**
    in Martian architecture 122, **242–243, 260**
    and microbes on Mars 158–159
    polar ice caps, on Mars **106–107**
    subsurface 30
Ice House (habitation) 122, **242–243, 260**
Immune system, changes in space 68
Impact craters **44–45**
In situ resource utilization (ISRU) 28–31
    fuel 121, 157
    life support fluids 121
    MOXIE 116
    water 30, 165
Indian Space Research Organization (ISRO) 24, 202, **214, 214–215**
Inflatables
    as habitats 123–124, **134–135, 224–225, 242–243**
    in landing 25, **42–43,** 53
    in reentry **142–143**
Innovative Advanced Concepts Program (NASA) 263
InSight (Interior Exploration using Seismic Investigations, Geodesy and Heat Transport) **184, 236–237**
International cooperation *see* Global vision
International Space Exploration Coordination Group (ISECG) 207–208
International Space Station (ISS)
    astronauts 36
    crew autonomy, lack of 75
    endurance mission 67
    Expedition 36/37 (2013) **64–65**
    growing food on 165
    Harmony module **86**
    international cooperation 204
    "ISSpresso" machine 91, **91**
    long-term space travel, research on 67, 71, 246–247
    as Mars analog 69, 71, **86, 88–89**
    modules 71
    and psychological stressors 87
    size 69
    Soyuz arrival **92–93**
    supply delivery **218–219, 232**
    3-D printing 121, **123**
    U.S. contribution to 205, 223
    zinnias grown on 67
ISECG (International Space Exploration Coordination Group) 207–208
Isolation, effects of 69, 71–75, 87
ISRO (Indian Space Research Organization) 24, 202, **214, 214–215**
ISRU *see* In situ resource utilization
ISS *see* International Space Station

J

Japan Aerospace Exploration Agency 202
Jet Propulsion Laboratory 28, 53, 125, 205, 207, **245**
John C. Stennis Space Center 39
Johnson, Gregory 125
Johnson Space Center, Texas 137, 157
Jupiter, moons of **180–181**

K

Kanas, Nick 87, **87**
Kazakhstan, space program **212–213**
Kelly, Mark 67–68, 97, **97**
Kelly, Scott
    ISS endurance mission 36, 67, **94–95, 96,** 97
    portrait **97**
    in Soyuz simulator **36–37**
    State of the Union address 203
    The Twins Study 67–68, 97
Kennedy, John F. **222**
Kennedy Space Center, Florida 28, 165, 203
Khoshnevis, Behrokh 122
Kononenko, Oleg **86**
Kornienko, Mikhail 36, 67, **94–95**
Kronos 1 **256–257**

L

Lambright, W. Henry 203
Landers, robotic 25
Landing sites, on Mars
    Curiosity rover 8–9, 54
    for human missions 25, 27, **29**
    Phoenix Mars lander 193, **209**
    Viking 2 lander 6–7
Langley Research Center 121, 246
Lansdorp, Bas 209–210
Laurini, Kathy 208
LavaHive 123–124, **260–261**
Lee, Pascal 72, 145, **145**
Levine, Joel 159
Lewis, Ruthan 246
Lichens **176**
Life, definition of 179
Life, on Mars
    drilling samples **190–191,** 193
    Earth analogs **152–153,** 162–164, **166–173, 186–187,** 193
    ethical issues 249, 251
    going green 165
    hazards to humans 163–164
    investigations into 54
    and landing site choices 27
    microbes 158–159
    nether regions 159, 161
    perchlorate problem 156–158
    risks to Earth from samples 156
    signs of 150–165
    and water 161–162
Lindgren, Kjell **86**
LIQUIFER Systems Group 123
    see also LavaHive
Lockheed Martin Space Systems **184,** 207, **226–227**
Logsdon, John 223, **223**
Long-term space travel, effects of
    ISS research 67, 71, 246–247
    The Twins Study 67–68
Love, Stanley 27, 118–119
Low-density supersonic decelerator **42–43**
Lowell, Percival 126
Lunar and Planetary Institute, Houston, Texas 25

M

Made In Space 121, 123
MAHLI (Curiosity rover's imager) 53
Malenchenko, Yuri 92

Mangalyaan (Mars Orbiter Mission) 24, 202, **214–215**
Manning, Rob 53, **53**
Manufacturing technologies, on Mars 121–122
Manzey, Dietrich 87
Maps
  Eastern Hemisphere North 6–7
  Eastern Hemisphere South 8–9
  potential exploration zones for human missions 58–59
  Western Hemisphere North 12–13
Mariner 9 Mars orbiter 158
Mars (National Geographic Channel series)
  background 16–17
  episode 1 24, **24**
  episode 2 68, **68**
  episode 3 116, **116**
  episode 4 156, **156**
  episode 5 202, **202**
  episode 6 246, **246**
Mars (planet) *see specific features, locations, missions, and topics*
Mars (planet), analogs
  Antarctica 71–72, 75, **76–81, 152–153**
  Devon Island 72–73, **144,** 145
  HI-SEAS 73–75, **82–85**
  ISS 69, **86, 88–89**
  Utah as **100–103**
Mars-1 rover 145
Mars 2020 mission 28, 116
Mars Analog Research Station project 72–73
Mars Architecture Steering Group 137
Mars Atmosphere and Volatile Evolution Mission (MAVEN) 24
Mars City Design competition 124–125
Mars Desert Research Station, Utah **100–103, 182–183**
Mars Express 24, **106–107**
Mars Global Surveyor 6–7
Mars Ice House 122, **242–243, 260**
Mars Institute 145
Mars Odyssey 24, **60–61, 108–109**
Mars One **182,** 209–210
Mars Orbiter Mission (Mangalyaan) 24, 202, **214–215**
Mars oxygen in situ resource utilization experiment (MOXIE) 116
Mars Reconnaissance Orbiter 24, 28, 116–117, 159, 161
Mars Science Laboratory spacecraft **20–21,** 25, **34–35,** 155
Mars Society 72–73, **100–103**
Mars World, Las Vegas, Nevada 254–255
Mars500 mission 69, **98–99, 209, 253**
Marsland (human settlement) 240–279
  anthropological critique 254
  habitations **242–243, 258–259**
  health concerns 245–247
  parks 251–252
  terraforming Mars 247–249, 251
  tourism poster **244**
  see also Habitations, on Mars
The Martian Landscape (Mutch) 255
Massachusetts Institute of Technology 115–116
Mauna Loa volcano, Hawaii 73–75, **82–83**
MAVEN (Mars Atmosphere and Volatile Evolution Mission) 24
McEwen, Alfred 116–117
McKay, Christopher **192,** 193, **193,** 248–249
Mental and emotional stresses, on Mars 62–75
  Antarctica as analog 71–72, 75, **76–81**
  crew conflict 74
  crew-ground disconnect 74–75
  Devon Island as analog 72–73
  "Earth out of view" phenomenon 87
  HI-SEAS as analog 73–75, **82–85**
  isolation 69, 71–75, 87
  ISS as analog 69, 71, **86, 88–96**
  NASA studies 87
  The Twins Study 68–69, 97
Meteorite impact craters **44–45**
Microbes
  carried to Mars 27, 158, 164
  producing oxygen on Mars 263
  search for, on Mars 158–159
  see also Planetary protection protocol
Microgravity 68
Microgravity Science Glovebox **123**
Migration to Mars *see* Habitations, on Mars; Human mission on Mars; Marsland
Mir space station 87, 204, **234–235**
MOA Architecture 124
Modi, Narendra **214**
MOM *see* Mars Orbiter Mission
Moon landing *see* Apollo missions
Moon-orbiting missions 207
Moon village, as precursor to human travel to Mars 137, 204–205, 207, 208, 223
MOXIE (Mars oxygen in situ resource utilization experiment) 116
Mueller, Robert 28–29, 165
Mulyani, Vera 124–125
Musk, Elon 16, 70, **200,** 210–211, 233
Mutch, Thomas A. "Tim" 255

N

Naica, Mexico
  cave of crystals **166–167**
NASA
  call for astronauts (2016) 211
  Evolvable Mars Campaign 233
  Human Research Program 73
  Innovative Advanced Concepts Program 263
  Office of Planetary Protection 185
  Regions of Interest 27
  see also specific facilities, missions, and vessels
NASA-ISRO Mars Working Group 202
National Additive Manufacturing Innovation Institute 122
National Space Biomedical Research Institute 247
National Space Society 247
NEO Native 124
New Shepard rocket **230–231**
New Space journal 205
Newman, Dava **264,** 265
Nielsen, Nick 253
Nili Patera (dune field), Mars **4**
Noctis Labyrinthus, Mars **108–109**
Nozomi (Planet-B) explorer 202
Nyberg, Karen **64–65**
Nye, Bill 160

O

Obama, Barack **200,** 203, 223
Ojha, Lujendra 159, 161
Okarian Humvee rover 145
O'Leary, Beth 251–252
Olympus Mons, Mars 12–13, 140, **140–141**
Opportunity rover 25, **112–113, 163, 190, 238–239**
Orbital ATK Cygnus cargo vehicle **92–93**
Orion spacecraft 32, **32–33,** 39, 137, 204, 207, **226–227**
Osteoporosis 246–247
Overview effect 252–254
Overview Institute 252
Oxygen, on Mars
  producing 115–116, **262,** 263
  as rocket fuel 116
  terraforming Mars 248–249

P

Panspermia theory 163
Parachute technologies **119,** 122–123
Parks, on Mars 251–252
Pass, Jim 105, **105**
Pathfinder/Sojourner rover 25
Perchlorates 115, 156–158, 161
Permafrost, as water resource 30
Phobos (Martian moon) **40–41,** 202, 205, 273
Phoenix Mars lander **22,** 25, **31,** 157, 193
Photosynthesis 248–249
Planet-B (Nozomi) explorer 202
Planetary protection protocol
  astronaut health concerns 163–164
  cross-contamination concerns 162
  in landing site selection 27
  for Mars-bound vehicles 158, 164, 174, **174–175**
  objectives 185
  responsibility for 185
  sacrifice zones 171
Planetary Science Institute 158
Planetary Society 160
Power Zone 117
Primary Lander Zone 117
Private sector efforts 209–211, **228–229,** 233
Prototype exploration spacesuit (PXS) **132,** 133

R

Radiation
  exposure on Mars 115
  risks 68, 245–246
  ultraviolet radiation 115
Reconnaissance craft, robotic 24, 28
Recurring slope lineae (RSL) 30, 159, **159,** 161–162, 165, **194–195**
Red Dragon (SpaceX capsule) 210
Reentry technology **142–143,** 143
Regions of Interest (NASA) 27
Regolith materials 123, 124, 134
Reiter, Thomas 204
Resources *see* In situ resource utilization
Retropropulsion, supersonic 25, 53
Robots and robotic crafts
  creating habitations 122–123, **266–267**
  delivering supplies **138–139**
  exploration of Mars 118, 171
  landers 25
  reconnaissance craft 24, 28
  robotic culture 252
  solar-powered **138–139**
Rock formations 163, **163**
Rosetta spacecraft, views from **18–19**
Rovers, nuclear-powered 155–156
Rovers, pressurized **125,** 145
RS-25 engines **38,** 39
RSL *see* Recurring slope lineae
Rummel, John 158
Rush, Andrew 121–122
Russia, space program 69, **98–99,** 205, **212–213**

S

Schiaparelli (entry-descent-landing module) **174–175,** 201
Schmitt, Harrison "Jack" 156
Science Applications International Corporation 117, 119
SEArch (Space Exploration Architecture) 122
SETI (Search for Extraterrestrial Intelligence) Institute 158
Settlement, on Mars *see* Habitations, on Mars; Human mission on Mars; Marsland
Shapiro, Jay 247
Sharp, Mount, Mars **114, 274–275**
"Sheepbed" rocks 163, **163**
Sky crane 49, **49**
Skylab 67
Slobodian, Rayna Elizabeth 254
SLS (space launch system) 39
Smith, Marcia 233, **233**
Smith, Peter 157
Smith, Scott 165
Snottite **170**
Snow White (ditch), Mars **31**
Social sciences 105
Soil analyses 165
Sojourner rover 25
Southwest Research Institute 162
Soyuz spacecraft **36–37, 92–93, 94–95**
Space and Technology Policy Group 233
Space Exploration Architecture (SEArch) 122
Space Exploration Technologies *see* SpaceX
"Space fog" 68
Space launch system (SLS) 39
Space Policy (journal) 251
Space race 31
Space radiation *see* Radiation
Space shuttles 39, **104,** 223, **234–235**
Space suits 119, **132–133,** 133, **264–265,** 265

Space Tourism Society 255
Space walks on Mars 118–119
SpacePolicyOnline.com 233
SpaceShipTwo **228–229**
SpaceX
Crew Dragon (craft) **220–221**
Dragon resupply craft **232**
government funding 233
hover test **218**
launches **200, 218–219**
multiplanetary goals 210–211
Spencer, John 255
Spennemann, Dirk 252
Spirit rover 25, 46, **46–47, 209, 276–277**
Stennis Space Center 39
Stickney Crater, Phobos **40–41**
Stillman, David 162
Subsurface ice 30
Super-greenhouse gases 248
Supersonic decelerator **42–43**
Supersonic retropropulsion 25, 53

T

Tardigrades 179, **179**
Team Gamma 122–123, 134
Techshot 263
Temperature, on Mars 115, 248
Terraforming Mars 171, 247–249, 251, 268, **268–269**
Tharsis Montes region, Mars 12–13
Thermophiles 176
Thibeault, Sheila 246
Thomas A. Mutch Memorial Station (Viking 1 lander) 155, 255
3-D printing
food 121
habitats 122, 123, 124, **134–135, 260**
space suits 133
tools and equipment 121, **123**
Time line, of Mars missions 280
Toups, Larry 117
"Tumbleweed craft" 118, **128–129**
Turner, Ron 245–246
The Twins Study 67–68, 97

U

Ultraviolet radiation 115
United Arab Emirates, space program 202–203, **216–217**
United Launch Alliance **226–227**
University College London, Centre for Altitude, Space, and Extreme Environment Medicine 247
University of Arizona 116–117, 157
University of Southern California, Center for Rapid Automated Fabrication Technologies 121–122
Urbina, Diego **98**
U.S. Senate Subcommittee on Science and Space 208–209
Utah, as Mars analog **100–103**

V

Valles Marineris, Mars
canals **127**
Coprates Chasma **194–195**
as habitation zone 124
Kronos 1 **256–257**
on map 14–15
Mars Odyssey images **60–61**
seasonal water budget 162
size 14
Victoria Crater, Mars **112–113, 238–239**
Viking landers 6–7, 25, 155, 164, 255
Villa Luz, Cueva de, Mexico **170**
Virgin Galactic SpaceShipTwo **228–229**
Vision, changes in 68
Volcanoes 12
Volkov, Sergey **94–95**
Von Braun, Wernher **222**
Voyager 1, images from 252

W

Wageningen UR 165
Wagner, Richard 72
Walking, on Mars 118–119
Wamelink, Wieger 165
Wastewater, recycling 121
Water, on Mars
aquifers 117, 159, 162, 164
brininess 30, 161–162, 165
evidence for 159, 161
as indicator of life 158–159, 161–164
landing site requirements 27
possible sources 30, 116–117
in situ resource utilization 29–30, 158
terraforming Mars 248
Western Hemisphere North, map 12–13
Western Hemisphere South, map 14–15
Wheeler, Ray 165
White, Frank 252–254
Why Mars (Lambright) 203
Wielders, Arno 209–210
Winds, on Mars 115, 118, **128–129**
Wood frogs **178–179,** 179
"Wopmay" rock 163, **163**
Wörner, Johann-Dietrich 204
Wray, James 161

Y

Yellowknife Bay, Mars 163, 193
Yellowstone National Park, U.S. **177**

Z

Z2 (space suit) 133, **133**
Zinnias **66**
Zubrin, Robert 72–73, 120
Zurek, Rich 28

Since 1888, the National Geographic Society has funded more than 12,000 research, exploration, and preservation projects around the world. National Geographic Partners distributes a portion of the funds it receives from your purchase to National Geographic Society to support programs including the conservation of animals and their habitats.

National Geographic Partners
1145 17th Street NW
Washington, DC 20036-4688 USA

Become a member of National Geographic and activate your benefits today at natgeo.com/jointoday.

For information about special discounts for bulk purchases, please contact National Geographic Books Special Sales: specialsales@natgeo.com

For rights or permissions inquiries,
please contact National Geographic Books Subsidiary Rights: bookrights@natgeo.com

Library of Congress Cataloging-in-Publication Data

Names: David, Leonard (Space journalist)
Title: Mars : our future on the Red Planet / Leonard David ; foreword by Ron Howard.
Description: Washington, D.C. : National Geographic, [2016] | Includes index.
Identifiers: LCCN 2016019897 | ISBN 9781426217586 (hardcover : alk. paper)
Subjects: LCSH: Space flight to Mars. | Mars (Planet)--Exploration. | Space colonies.
Classification: LCC TL799.M3 D38 2016 | DDC 919.9/2304--dc23
LC record available at https://urldefense.proofpoint.com/v2/url?u=https-3A__lccn.loc.gov_2016019897&d=DQIFAg&c=uw6TLu4hwhHdiGJOgwcWD4AjKQx6zvFcGEsbfiY9-El&r=Ar3XRLWsOd9X4qagesooQpv_FSetDc1lkl9pxdlLrhw&m=D-JEF77Qj8hCHNJPP_AaXCmEAvHqPkSTFtvQs86STVUw&s=SBpCuK58C6ejTlls2VJJmvMVlNOEXSaTqTVu1Y19JDc&e=

Printed in the United States of America

16/QGT-RRDML/1